The Second Quantum Revolution

Volume 2

Foundational Transformation of Quantum Optics

AQM Theory of the Photon

Victor Vaguine, Ph.D.

ConsReality® Press

ISBN: 978-1-936795-20-8
Library of Congress Control Number: 2020939603

Published in the United States of America
ConsReality® Press.
Dallas, Texas
www.consreality.com

Publisher's Cataloging-in-Publication data

Victor Vaguine. Ph.D.
The Second Quantum Revolution.
Volume 2
Foundational transformation of quantum optics.
AQM theory of the photon / Vaguine, Victor

ISBN: 978-1-936795-20-8
Includes bibliographical references and Index

1. Quantum optics 2. Physics 3. Philosophy 4. Experiment. I. Title

QC446.3 .V-34 2020
535. 15 - dc22 2020939603

Editing by Galena Miller-Horii.
Cover design by Andre Vaguine / ConsReality, Inc.
Illustrations by Andre Vaguine / ConsReality, Inc.
Book design and production by Andre Vaguine.

Edition 1.0

To Vera.

Acknowledgements

My sincere and profound appreciation to my son Andre Vaguine for his dedication and unwavering support.

“All these fifty years of conscious brooding have brought me no nearer to the answer to the question ‘What is light quanta?’ Nowadays every physicist thinks he knows it, but he is mistaken.” (Albert Einstein, Letter to M. Besso, 1951).

CONTENTS

List of Postulates – Volume 2
(continued from Volume 1)

Postulate 41.	No Pathway Exists for Photon Traveling in Free Space to Spontaneously Convert Itself Into Another Inner Structure
Postulate 42.	Detailed Energy Transformation Process of Photon Interaction in Optical Domain including Detector
Postulate 43.	The Position Parameter Distribution in the Large Assembly of Monoenergetic Photons
Postulate 44.	The Complete Definition and Properties of the Photon based on Aphysical Quantum Mechanics

Prologue to Volume Two

Science is the uncompromising search for the truth in objective reality.

No force in the world can stop fundamental scientific truth.

For twelve years, beginning in 2008, I have been working diligently, as an independent scientist outside of the scientific establishment, on problems of foundational physics developing my Aphysical Quantum Mechanics (AQM), a deeper quantum theory. The result of my scientific work is published in three separate volumes under the title The Second Quantum Revolution [1], [2], [3]:

- Volume 1: Foundational Transformation of Quantum Mechanics. Aphysical Quantum Mechanics as Deeper Quantum Theory. Elementary Consciousness of Elementary Particles.
- Volume 2: Foundational Transformation of Quantum Optics. AQM Theory of the Photon.
- Volume 3: Foundational Transformation of Quantum Electrodynamics. Two Branches of Electromagnetism.

Some might say that my statement about the Second Quantum Revolution is an exaggeration.

In fact, it is an understatement!

Aphysical Quantum Mechanics (AQM) is a deeper and more profound quantum theory. In principle, one would not expect AQM to produce immediate scientific revolutions in other branches of fundamental physics, such as quantum optics and elementary particle physics, but that is exactly what has happened. Both of these branches—quantum optics and especially elementary particle physics—have undergone the dramatic foundational transformation.

Aphysical Quantum Mechanics is the origin of the Second Quantum Revolution.

At the Fifth Solvay Conference in 1927, the Copenhagen Group declared that the foundation of Quantum Mechanics (QM) was completed.

This was the First Quantum Revolution. It was an outstanding scientific achievement. It opened a window into quantum reality.

However, since 1927, for almost 100 years, QM has remained a frozen monument to quantum dogmatism surrounded by the deficient paradigmatic perimeter. QM has never been completed. Major pieces of the puzzle are missing.

There have been multiple attempts by many quantum physicists to explain QM on a basis of numerous QM interpretations. Predictably, all those interpretations have failed because these attempts have been made within the deficient paradigmatic perimeter of the QM dogma.

There have been also numerous attempts to get through the perimeter with *single issues*, such as search for the electron three-dimensional inner structure or exploring role of consciousness in quantum reality. These efforts have failed as well.

These numerous QM interpretations have never been able to penetrate through the deficient paradigmatic perimeter of quantum mechanics. Furthermore, a break through in a single point of the perimeter is not sufficient. It requires the critical mass of foundational ideas and concepts in order to arrive at the Second Quantum Revolution. Incremental progress and incremental breakthroughs are not sufficient.

My approach is different. I have the advantage of being an independent scientist outside of the scientific establishment. During my 12 years of scientific work, I did not need to endure the overwhelming and depressing quantum dogmatism and the imposing groupthink pressure. I proceeded in a systematic way to destroy the QM deficient paradigmatic perimeter with massive blows directed at multiple points, by applying the mass of fundamental concepts in the form of 30 postulates in Volume 1, 14 postulates in Volume 2 and 31 postulates in Volume 3. Those fundamental concepts are not conflicting but rather reinforcing each other. As a result, the paradigmatic perimeter has been crushed, and *a deeper and more profound quantum reality has shown up.*

The conclusion is that the first quantum revolution has never been completed. For all those decades since 1927, QM has been in a suspended state, seeding confusion, absurdities, and fundamental misconceptions.

My scientific work sets in motion the second quantum revolution, the genuine and profound one. It brings at the outset an avalanche of fundamental concepts and scientific discoveries.

AQM dramatically expands fundamental understanding of quantum reality by postulating, in addition to the physical category, two more fundamental categories: the aphysical category and the elementary consciousness of elementary particles. There is no way one can comprehend the quantum reality without these two additional fundamental categories.

I discovered and explained three-dimensional inner structures and self-masses for each class of elementary particles and their individual interactions in spacetime dynamics, which can be visualized in the human mind at every step. New principles and new particles (in fact, new classes of elementary particles!) are postulated, with a plethora of new fundamental properties.

It turned out that "the bread and butter particles", such as the electron, the photon, the neutrino, the proton, and the neutron, have been left behind and, surprisingly, fundamentally hardly explored. Whatever we knew about these particles prior to the Second Quantum Revolution was only the tip of the iceberg.

Volume 2 is devoted to the photon. Therefore, it is appropriate to bring upfront the direct comparison between the QM photon and the AQM photon as the confirmation of the above "bread and butter" statement. It is done in the form of Addendum to Prologue for Volume Two.

The paradigmatic power of Aphysical Quantum Mechanics is such that I was able to uncover 53 fundamental misconceptions/absurdities in the Standard Model theory of particle physics; and to make 27 fundamental scientific discoveries presented in all three Volumes.

AQM opens the floodgates of new physics. It brings democracy into foundational physics. Graduates and undergraduates will be able to select new physics problems to work on among the thousands immediately available. Hundreds of doctoral dissertations will be written based on AQM. And most importantly, for a change, the general public will be brought along.

In his book, "*The Structure of Scientific Revolutions*", [4] Thomas S. Kuhn observes that genuine scientific revolutions are manifested by the destructive-constructive paradigm, meaning the massive destruction of old information and its replacement by new.

I have to state that not all pre-revolution theories and concepts are destroyed. Those that are along the natural scientific trajectory toward a deeper understanding of objective reality survive and flourish.

That exactly what is happening in the Second Quantum Revolution driven by Aphysical Quantum Mechanics (AQM). It is not by chance the Second Quantum Revolution is originated outside the scientific establishment.

The Copenhagen and other QM interpretations are destroyed and replaced by a new AQM paradigm. All known long-lasting quantum enigmas, paradoxes, and mysteries are resolved. As far as QM is concerned, its role is about to shift from foundational physics to a mathematical quantum application tool, to be used for calculations of statistical values with some paradigmatic limitations recognized and applied.

The concept of aphysical energy is one of my fundamental discoveries.

Here is a brief explanation, especially important for the discussion of the photon :

Physical energy and aphysical energy are two intertwined fundamental entities. They operate within the spacetime dynamics in proportionality. One cannot function without the other. Physical spin coexists with aphysical spin. Physical angular momentum coexists with aphysical angular momentum. Physical kinetic momentum coexists with aphysical kinetic momentum. Physical charge coexists with aphysical "imitation" charge. All of them coexist in strict proportionality within the energy conservation law and in strict compliance with the universal constant U.

It would be an oversimplification to think that aphysical energy properties are always aligned with physical energy properties. Aphysical energy has its own set of unique properties, such as *entanglement, self-entanglement, and instantaneous action at any distance. Aphysical energy brings aphysical determinism into quantum interactions. Aphysical energy has no inertial or gravitational properties.*

Physical energy is responsible for physical properties of elementary particles and can be described to a significant extent in a classical way. *Surprisingly, physical energy plays no role in interference or diffraction. It is aphysical energy that is responsible for interference and diffraction.* Physical energy allows us to detect and register interference and diffraction patterns produced by aphysical energy. The role of aphysical energy in the photon self-entanglement is massive.

Elementary consciousness is responsible for the origin of spin and for "the collapse of the quantum state vector". "The collapse" has no ontological explanation or meaning. In my scientific terminology, elementary consciousness is responsible for the instantaneous reconstruction of the self-entangled quantum state into the full-fledged quantum state. In addition, elementary consciousness is responsible for entanglement, self-entanglement, non-locality, and instantaneous action. I foresee strong opposition to the concept of elementary consciousness of non-living matter, but a genuine scientific revolution should expect strong resistance from the status quo defenders.

Elementary consciousness of non-living matter is one of AQM's principal concepts. Elementary consciousness of elementary particles should be understood in totality of its specific properties. Nothing more, nothing less. It is not an excuse, as in the cases of the Schrödinger cat or Many Worlds interpretation, for endless and mindless speculations, brainwashing the general public for many decades.

Progress in fundamental physics is based upon the balance of intuition and mathematical logic. We currently have a paradoxical situation where mathematical logic is predominant and excessive in its virtuosity at the expense of physics. The ample reservoir of intuition in foundational physics remains unused. My work is based predominantly on intuition (in fact, *super intuition*), which, as Einstein states, is more important than mathematical logic.

Here are some principal points of the quantum reality according to Aphysical Quantum Mechanics:

- Quantum reality consists of three fundamental categories: physical, aphysical, and elementary consciousness.
- One of the core principles of AQM is spacetime visualization of individual elementary particles and individual elementary quantum processes. *Physics without spacetime visualization has no future.*
- All elementary particles have non-zero size. All elementary particles have the three-dimensional inner structure with perfect geometry consisting of *the physical energy c-ring(s), the aphysical energy cylinder(s), and elementary consciousness.*
- The three-dimensional inner structure with perfect geometry is in conflict with the Heisenberg uncertainty principle, which is a probabilistic principle, not relevant for individual quantum entities and/or quantum individual processes. Its dogmatic interpretation has been one of major roadblocks on the road toward progress in fundamental quantum physics.
- The inner structure of the fundamental elementary particle has a single c-ring and a single aphysical cylinder; the inner structure of the composite particle has two or more c-rings and two or more aphysical cylinders.
- Spin is the expression of elementary consciousness. Spin is the actual rotation in three-dimensional space at Compton angular velocity. Spin is the mode of existence of the elementary parti-

cles and is eternal over their lifetimes. There are no elementary particles without spin, obvious or hidden. Spin can be visualized in a classical way. The elementary consciousness is the force driving spin. No elementary consciousness – no spin – no self-mass.

- Elementary particles are particles, never waves. Wave property is a probabilistic expression only. It is well known that a detection screen always shows only individual dots.
- All elementary particles in free space have trajectories. The full-fledged elementary particle has a single trajectory. The self-entangled elementary particle has several trajectories but only one carries the physical substance. All other trajectories are aphysical.
- The elementary particle has a well-defined position in its frame of self-reference with zero momentum. This is also valid for bosons in their frames of self-reference v = c. The origin of self-mass ("self-energy") for fermions and bosons (yes, for bosons!) is established and explained. Both fermions and bosons have self-mass. There is no more division of "massles" and "massive". Yes, bosons are no longer "massles". Bosons have self-mass and zero kinetic energy in their frames of self-reference, v = c. Virtual energy does not exist.
- A self-entangled elementary particle can be in several places at the same time, physically always in one and aphysically in all the others. Just prior to physical-physical interaction, a particle is instantaneously reconstructed to the full-fledged quantum state regardless of distances separating the host and a-fractions, with no violation of special relativity.
- After physical-physical interaction of one elementary particle with another elementary particle or with a macroscopic entity, such as a measurement device, no entanglement between them

remains in the aftermath of their interaction. Without much ado, the perennial "measurement problem " exists no longer. There is an exception under some special conditions such as BE condensation at microKelvin temperature.

- AQM has expanded understanding of determinism and causality in the description of the quantum world. Yes, indeed, there are "hidden variables", such as position parameters. For example, no two muons are identical. The difference in their lifetimes is explained by the difference in initial values of their position parameters. *A decay of the individual particle is the deterministic process.*
- AQM explains that there are at least two fundamental sources of the probabilistic nature of the quantum reality such as (1) irreducible quantum randomness in the selection of the host in self-entanglement; and (2) irreducible quantum randomness in the position of the physical c-ring relative to the aphysical cylinder in the inner structure of the elementary particle.
- In the world of elementary particles, there exists a harmonious coexistence of the classicality and the quantum. The classical-quantum divide does not exist. *In AQM, the classicality has found its perfect expression.*
- Quantum laws are not universal. The quantum reality is a subset of the objective reality. Gravity is not part of the quantum reality. Humans and cats are not a subset of quantum reality.

A new challenge for the mathematically inclined physicists is to develop a mathematical formalism for the description of individual elementary quantum processes in spacetime dynamics. This is what is on the horizon and coming to quantum physics in the near future. However, even without new mathematics, I was able to reconstruct in all details the spacetime dynamics of selected individual elementary quantum interactions and processes, such as the muon decay and the subsequent formation of the electron.

In my firm view, the consciousness is the most important and enormously significant fundamental category. However, there is very little understanding of the consciousness as a fundamental category of objective reality. Consciousness cannot be derived from anything and neither can it be reduced to anything.

With respect to the foundational transformation of quantum optics, here are some principal points presented in Volume 2:

- New photon properties are discovered and explained, such as the photon three-dimensional physical-aphysical inner structure. A special case is the three-dimensional physical-aphysical inner structure of the linear polarized photon, with "hidden" spin consisting of two half-spins opposing each other.
- I discovered and introduced the concept of *self-entanglement.*
- In case of the photon, self-entanglement plays the most important role in photon interference and diffraction. Compared with entanglement, which is delicate and rare, self-entanglement is the massive phenomenon in quantum optics.
- The photon inner structure consists of the physical energy c-ring, the aphysical energy cylinder, and elementary consciousness residing in the c-ring (see Figure 1-P).
    - I declare that the interaction of optical photons with ideal optical materials/components, including ideal crystals and excluding detectors and other absorbing materials, is *not physical but aphysical.*
    - Based on the discovered properties of the photon inner structure, I concluded that the cross-section of photon-

photon interaction is zero. This applies to all bosons. Boson-boson interaction does not exist in any variations and under any circumstances.

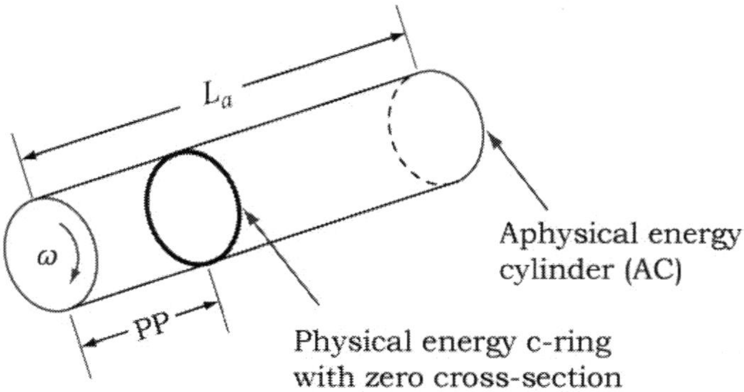

Figure 1-P.
Three-dimensional inner structure of the circularly polarized photon. PP – position parameter.

- The photon aphysical length can be measured experimentally in quantum optics research laboratories. The aphysical length, normalized by photon wavelength, is *the photon fundamental aphysical constant.* It is one of the first aphysical fundamental constants in particle physics to be determined experimentally with better accuracy in the near future. The photon aphysical constant is as important to foundational physics as the Planck constant, the speed of light, and the electric charge. In the world of elementary particles, there are other aphysical constants. The photon "wavelength" means the length of the circumference of the aphysical cylinder.
- Figure 2-P shows AQM prediction versus QM expectation. I was able to derive only a single experimental point from available and unrelated experimental data. We can all finally agree that

the saying “QM has never been proven wrong” is a myth. It has always been a myth.

- The Heisenberg uncertainty principle and the wave function represent mental blocks for quantum foundational physicists. I have illustrated this using individual photons.
- According to AQM and as shown in Figure 2-P, the more precisely we measure photon energy, the more precisely we determine the photon aphysical length.
- With all of its new properties discovered, the photon turns out to be one of the most versatile elementary particles.

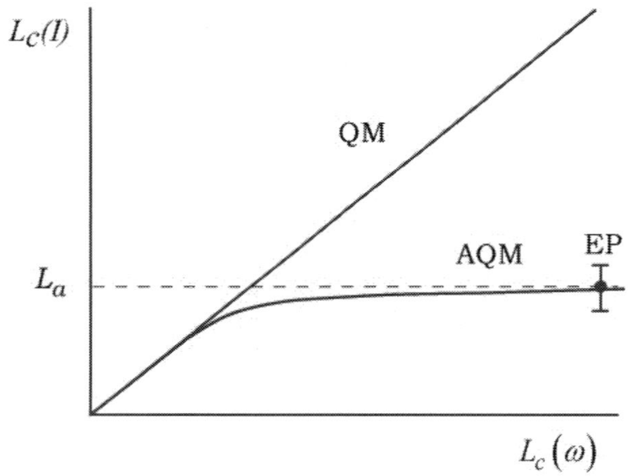

Figure 2-P.

Expectation of Quantum Mechanics (QM) versus prediction of Aphysical Quantum Mechanics (AQM).

Explanation to Figure 2-P:

λ_c - photon wavelength, which is equal to the circumference of the aphysical cylinder;

$L_c(I)$ – photon interference pattern length; $L_c(\omega)$ – coherence length of photon source, same as the length of photon wave packet; L_a – photon aphysical length; L_a / λ_c – photon fundamental aphysical constant

I was able to derive only a single experimental point EP from available and unrelated experimental data (see Chapter 8).

Example: for λ_c equal to 0.6 μm, experimentally estimated photon aphysical length is equal to 120 μm (+100 μm, -30 μm).

♦ ♦ ♦

Postscript: By publishing *The Second Quantum Revolution* I have fulfilled my promise given in my book *Prologue to Super Quantum Mechanics* (2012) [5].

♦ ♦ ♦

Addendum to Prologue for Volume Two

Comparison of the Photon in SM and AQM Interpretations.

A. The photon in Standard Model ("the SM photon")

The SM photon:

- is an elementary particle;
- is the quantum of electromagnetism;
- is the force of electromagnetism;
- is the mediator of electromagnetic interaction;
- is boson traveling in free space with the speed of light and is stable;
- is massless;
- has spin ± 1 in $\hbar$ units;
- has zero size;
- has no inner structure;
- has electromagnetic energy $E_P = \hbar\omega$; and
- has particle-wave properties.

In quantum optics which is an extension of SM, the photon is a wave packet and has spin in the range [-1, +1] in $\hbar$ units.

As identified in Chapter 3, in quantum optics experiments with a single-photon source, there are nine unexplained enigmas. For example, there is no ontological explanation on how single photon with the circular polarization is transformed into photon with the linear polarization.

B. The photon in Aphysical Quantum Mechanics ("the AQM photon")

The complete definition and properties of the AQM photon are presented in Postulate 44.

The photon can be understood only on the basis of the AQM quantum reality, which includes three fundamental categories: physical, aphysical, and elementary consciousness.

The AQM photon:

- is an elementary particle;
- is the quantum of the *inverted* electromagnetic energy;
- is the force of electromagnetism;
- is the mediator of the electromagnetic interaction;
- is boson traveling in free space with the speed of light, and is stable;
- as boson, it carries no charge;
- is a particle and never a wave;
- has spin ± 1 in $\hbar$ units;
- has spin in the range of [-1, +1] in $\hbar$ units in quantum optics;

- has the inner structure consisting of three constituents: the physical energy c-ring, the aphysical energy cylinder, and the elementary consciousness residing in the c-ring; the inner structure is shown in Fig. 1-P;
- has photon physical energy $E_P = \hbar\omega$ and aphysical energy $E_a = E_P / U$, where U is the universal constant;
- has physical self-mass $h\omega / c^2$; the origin of the physical self-mass is spinning of the c-ring; the driving force of c-ring spin and aphysical cylinder spin is the elementary consciousness residing in the c-ring;
- the normalized aphysical cylinder length L_a / λ_c is a photon fundamental aphysical constant; its value is found experimentally in approximation;
- the photon and the electromagnetic field are two fundamentally different categories; the origin of the photon *is not* the electromagnetic field; the c-ring is the bundle of *the inverted electromagnetic energy* spinning with Compton angular velocity at Compton radius.

C. Some more about AQM photon and its properties

In the optical energy range, the c-ring physical energy bundle can be split into two sub-bundles in any ratio, each spinning with equal but opposite Compton angular velocity, thus explaining spin values in the range of [-1, +1]. In case the physical energy bundle is split equally, spin is also split in two equal sub-spins opposing each other and resulting in zero spin ("hidden spin"). In such a case, photon acquires the linear polarization. This explains enigma of a transformation of a single photon from the circular polarization into the linear polarization.

The photon has neither momentum nor kinetic energy in any frames of references, including its own frame of self-reference (v = c). It is a counterintuitive statement.

The photon "inverted" electromagnetic energy is contained in the c-ring as a physical energy bundle, with uniform distribution of energy density along the circumference. In compliance with special relativity law, the c-ring cross-section is zero.

The photon's inner structure spins around its axis at Compton angular velocity $\omega = c / r$, where r is Compton radius. Zero cross-section of the c-ring means that the cross-section of photon-photon interaction is zero. This explains why the universe is to a significant extent transparent to traveling photons.

The position of the c-ring along the length of the aphysical cylinder is defined by the position parameter, PP, which is the irreducible quantum random parameter. For large assembly, the statistical distribution of PP along the length of the aphysical energy cylinder is uniform. The c-ring physical energy plays no role in interference or diffraction.

The spinning photon c-ring has enormous centrifugal force without centripetal force for balance.

The following is an explanation:

Traveling in free space, photon is precluded from expansion and release of its physical energy because simply there is no pathway available for such release.

Pair of photons can be entangled via c-link connection.

In the aphysical energy domain, photon acquires self-entanglement. The aphysical domain, such as the no-loss optical material, is the domain of aphysical-aphysical interaction only. In the aphysical domain, photon aphysical energy cylinder splits on any number of a-fractions, with only one, the host, carrying c-ring with physical energy. Each a-fraction is connected to the host by its individual c-link. This quantum state is called self-entanglement. It is the massive phenomenon explaining photon interference and diffraction. In the case of self-entanglement, a-fractions and the host travel independently along their individual trajectories and participate independently in elastic aphysical–aphysical interactions with other quantum entities.

In the optical aphysical domain, in addition to primary self-entanglement, individual a-fractions experience higher order of self-entanglement via aphysical-aphysical interactions of individual a-fractions with the aphysical structure.

In certain relationship between periodicity of multi-slit structure and photon wavelength, interaction of the full-fledged photon with such multi-slit structure results in photon self-entanglement, where the host with intact physical c-ring travels through one slit and the other a-fractions travel through the other slits. The photon wavelength is a misnomer. What is called photon wavelength is the length of the circumference of the photon aphysical cylindrical inner structure. Photon wave property is the statistical expression of assembly of monoenergetic self-entangled photons interacting aphysically with periodic aphysical structures such as multi-slits or crystals.

The interaction of optical photons with ideal optical material is not a physical process. Such process is governed exclusively by aphysical energy via aphysical-aphysical interaction. In a "no physical energy transfer environment", such as transparent optical materials in quantum optics, it is the aphysical substance of the photon, in the form of a set of a-fractions including the host, that plays the role in dynamics of quantum optics. The aphysical substance is the deterministic player in quantum optics, and therefore quantum optics should be called 'aphysical quantum optics.'

If an individual a-fraction of self-entangled photon attempts to engage in aphysical-physical interaction with another quantum entity, it is immediately recalled by the host. If the host meets such physical quantum entity, the host instantaneously recalls all a-fractions and transforms itself into the full-fledged photon. This is the explanation of the so-called "collapse of wave function."

In case photon encounters an absorbing quantum entity such as detector, it proceeds toward its transformation via physical-physical and physical-aphysical interactions with detector. Photon converts its physical and aphysical spinning energies into kinetic energies and transfers its kinetic energies to the absorbing entity together with physical and aphysical spins. Photon is absorbed and destroyed with no resulting entanglement.

References for Prologue to Volume Two

[1] Victor Vaguine, *The Second Quantum Revolution. Volume One. Foundational Transformation of Quantum Mechanics. Aphysical Quantum Mechanics as Deeper Quantum Theory. Elementary Consciousness of Elementary Particles.* ConsReality Press, 2020.

[2] Victor Vaguine, *The Second Quantum Revolution. Volume Two. Foundational Transformation of Quantum Optics. AQM Theory of the Photon.* ConsReality Press, 2020.

[3] Victor Vaguine, *The Second Quantum Revolution. Volume Three. Foundational Transformation of Quantum Electrodynamics. Two Branches of Electromagnetism.* ConsReality Press, 2020.

[4] Thomas S. Kuhn, *The Structure of Scientific Revolutions,* third edition, The University of Chicago Press, 1996.

[5] Victor Vaguine, *Prologue to Super Quantum Mechanics.* ConsReality Press, 2012.

Chapter 1

Introduction - Einstein and Quanta.
Historical background.

Quantum mechanics (QM) is the foundation of the Standard Model theory of particle physics (SM) which states that elementary particles are point-like, with no inner structure, with spin assigned as a quantum parameter with no classical analogy applied. The elementary particle spin cannot be visualized.

That is the mindset of quantum physicists today. It is a foundational trap.

Numerous attempts by many physicists to reverse such a situation by pursuing other approaches have not been successful.

The subject of this Chapter is the photon, one of the earliest discovered elementary particles. It was Einstein who in his 1905 paper put forward the hypothesis that light itself is quantized in packets of energy. He called those packets "light quanta." It was a profound observation.

In the 1909 paper, Einstein made another scientific contribution which stated that light quanta have both particle and wave properties. Einstein's theory successfully explained photoelectric effect. Einstein made these fundamental contributions to the foundation of quantum mechanics in its early stage of development.

For many decades, the photon has been scientifically studied in quantum optics research laboratories and high energy physics centers around the world. Thousands of scientific articles and hundreds of books have been published about the photon. Whole sectors of the economy and several high technology industries are based on the photon and its properties. It is difficult to find a high technology device or instrumentation which does not include, in some way, an application of the photon. But the scientific story of the photon is not yet over.

In Einstein's view, the photon complexity was (and is) underestimated. What QM and quantum optics tell us about the properties of the photon is only the tip of the iceberg.

Aphysical quantum mechanics (AQM) opens a new chapter in the understanding of photon fundamentals. AQM represents a genuine scientific revolution in quantum optics.

Chapter 2

The Photon in Standard Model Definition

According to the Standard Model theory of particle physics (SM), the photon as an elementary particle is point-like, with no inner structure, and with spin assigned by proclamation as an intrinsic quantum parameter. SM emphasizes that quantum spin has no classical analogy. The photon belongs to a class of gauge bosons. It carries no charge and is stable. The photon is the quantum of the electromagnetic field and the force carrier for electromagnetism. The photon exhibits wave-particle duality, and as a massless relativistic particle, it travels with the speed of light in free space

The energy of photon E is proportional to its frequency ν and inversely proportional to its wavelength λ:

$$E = h\nu = \hbar\omega = hc/\lambda ,$$

where ω is the angular frequency and $\lambda = c/\nu$. Photon spin S is energy (and frequency) independent.

In the context of this study, the most important photon property to consider is *spin*. According to SM, the photon as a gauge boson for electromagnetic interactions has two quantum spin states S=±1 (in $\hbar$ units), corresponding to the classical concepts of right-handed circular polarization with spin in the direction of travel, and left-handed circular polarization with spin opposite to direction of travel.

In the classical Maxwell electromagnetic theory, the linear polarization of an electromagnetic wave can be presented as the superposition of right- and left-handed circularly polarized electromagnetic waves of equal amplitudes. For photon, such approach is a misconception. The linear polarization cannot be explained on the basis of photon assembly. By using standard optical components, experimental physicists can convert individual photons into any quantum spin states in a range [1, -1].

However, probabilistic quantum mechanics, which is the foundation of SM, provides no ontological explanation for the linear polarization of an individual photon. It is yet another enigma, one among many.

Due to the deficiency of QM paradigm, a plethora of photon properties is overlooked in SM and quantum optics. Those new photon properties are discovered and explained in this volume on the basis of AQM.

The photon is one of the most important elementary particles in its application in science, research, industry, and economy. The photon is hardly explored while gigantic effort is directed toward research of exotic, high order elementary particles at ever higher energy.

Chapter 3

Quantum Optics Beam-Splitter Experiment with Linearly Polarized Photons in Quantum Mechanics Interpretation. Quantum Enigmas

3.1 Introductory Comments

Einstein stated: "…if the statistical quantum theory does not pretend to describe the individual system (and its development in time) completely, it appears unavoidable to look elsewhere for a complete description of the individual system…" [1].

Following Einstein, in order to demonstrate fundamental limitations and deficiencies of QM paradigm, one can *conceptually* analyze experiment performance and experimental data dealing with individual elementary quantum entities and elementary quantum processes. The most promising field for accomplishing this is quantum optics. As compared to particle physics, quantum optics experiments are much less expensive. In quantum optics, one can use recently developed single-photon source, the down-conversion (to be elaborated later) which delivers one photon at-a-time.

Here I propose to use the simplest conceptual quantum optics experiment, which includes the down conversion as a single-photon source of linearly polarized photons, and the beam splitter. Such experiment has been performed routinely for many years, even decades, in many quantum research labs, including those for college undergraduates. I propose to proceed step by step with a conceptual analysis of this experiment in both QM interpretation (see Chapter 3) and AQM interpretation (see Chapter 5).

The QM analysis is concluded and published in detail in my paper "Photon Non-physical Interactions and Quantum Enigmas" [2] a few years before I completed the Aphysical Quantum Mechanics. In this simple experiment, I observed nine quantum enigmas and a number of fundamental contradictions. The experiment performance indicates the existence of photon *non-physical interactions.* As stated in my paper, "All this is a strong indication that the QM paradigm is deficient. Something is missing in our understanding of fundamentals of quantum reality. As a probabilistic theory, QM, in principle, cannot explain the ontology of individual elementary quantum entities or individual elementary quantum processes. If we understand this, we would be able to proceed in a major way toward the Second Quantum Revolution."

3.2. Understanding of Photon in Quantum Optics

In quantum optics, it is customary to present photon as a wave-packet with all its physical parameters, such as angular frequency ω, wavelength λ, energy $E = \hbar\omega$, momentum $p = E / c$, and spin S. In quantum optics, a value of photon spin depends on degree of the photon polarization. As experimentally shown, spin can have any value in the range of $[-1,+1]$ in $\hbar$ units. At minimum polarization, called the linear polarization, spin is zero. Such range of photon spin values is unique only to quantum optics, where photon can have the circular polarization, the linear polarization, or the elliptical polarization, as demonstrated experimentally.

The reason is that in optical energy range there are optical transparent materials with no photon energy losses or photon energy transfer. One cannot find such materials in X-ray energy range or especially in high energy physics. In those fields, the photon can be brought to only two spin states, $S = \pm 1$.

What is a photon wave-packet? Is it a probabilistic mathematical entity or a physical object? The wave-packet cannot be a physical object since a photon as an elementary particle traveling in free space with velocity of light, in accordance with special relativity, has no dimension in the direction of travel.

This is Enigma 1.

How is it possible for an individual photon, with no physical dimension in the direction of travel, to interfere with itself?

This is Enigma 2.

3.3. Beam-splitter experiment with linearly polarized photons

An experimental layout is shown in Figure 1. It is simplified but sufficient for our purpose. A more elaborate one is discussed later in another conceptual experiment in conjunction with Mach-Zehnder interferometer.

A linearly polarized photon is sent to a polarizing beam splitter (PBS) at 45 degrees between photon plane polarization and PBS horizontal plane. The question arises—why can't QM explain ontology and visualization of an individual linearly polarized photon?

This is Enigma 3.

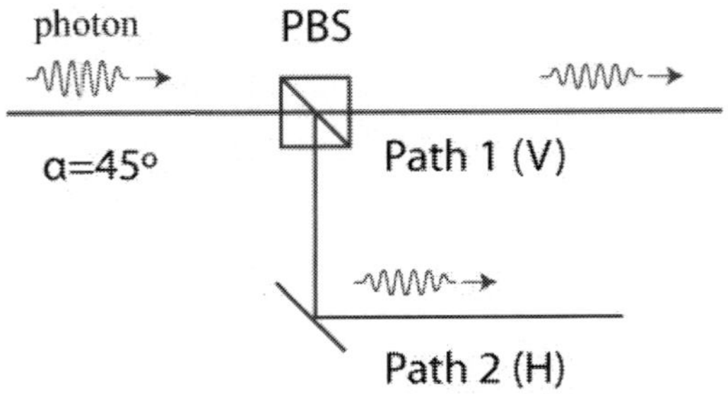

Figure 1.
Beam-splitter experiment with a source of linearly polarized photons in QM interpretation (one photon-at-a-time)

After passing through the PBS, photon is in the superposition of two quantum states with vertical linear polarization (V) on Path 1 and horizontal linear polarization (H) on Path 2 (or vice versa). It is described by wave function:

$$\Psi_0 = \frac{1}{\sqrt{2}}\Psi_1(V) + \frac{1}{\sqrt{2}}\Psi_2(H)$$

Photon has equal probability to be found on Path 1 or Path 2. According to QM, photon is in both places at the same time.

How is it possible for a physically indivisible photon to travel along two separate paths at the same time?

This is Enigma 4.

How can one explain the ontology of superposition of the linear polarization? QM cannot explain it.

This is Enigma 5.

3.4. The experiment interpretation on the basis of quantum positivism

Quantum positivism, as expressed by Bohr [3], Heisenberg [4], and other members of the Copenhagen Group, denies "any form of physical reality to the dynamic properties (such as position, velocity, momentum, energy) of a quantum system unless they are actually measured" [5]. Quantum positivists state that a quantum particle is in intrinsic fuzziness in all potentialities; only when measured or observed it arrives at a definite quantum state.

Here is an example of a quantum positivism explanation of our beam-splitter experiment where a single photon is in two places, Path 1 and Path 2, at the same time. To paraphrase Brian Green's eloquent and "pure" quantum positivism formulation [6], which I adapted to the beam-splitter experiment: *After emerging from the beam splitter, a photon hovers in quantum limbo in a fuzzy, amorphous, probabilistic mixture of two possibilities; only when measured one definite outcome emerges from two possible quantum states: either Path 1 with the vertical polarization or Path 2 with the horizontal polarization.*

According to quantum positivism, a photon is not real unless it is observed or measured.

Einstein stated: "I am not a positivist. Positivism states that what cannot be observed does not exist. The conception is scientifically indefensible, for it is impossible to make valid affirmations of what people 'can' or 'cannot' observe. One would have to say that 'only what we observe exists' which is obviously false." [7]

3.5. Interpretation of the experiment by quantum optics physicists

Let us further discuss the experiment of the previous layout (Figure 1) now with two detectors D_1 and D_2 installed, as shown in Figure 2. Probabilistically, only one half of photons pass through the PBS as vertically polarized and registered by detector D_1, and the other half get reflected as horizontally polarized and registered by detector D_2. Each photon is registered with equal probability 50/50 by only one of the two detectors. The question quantum optics physicists ask: *when* does an individual photon decide which path to take, Path 1 or Path 2?

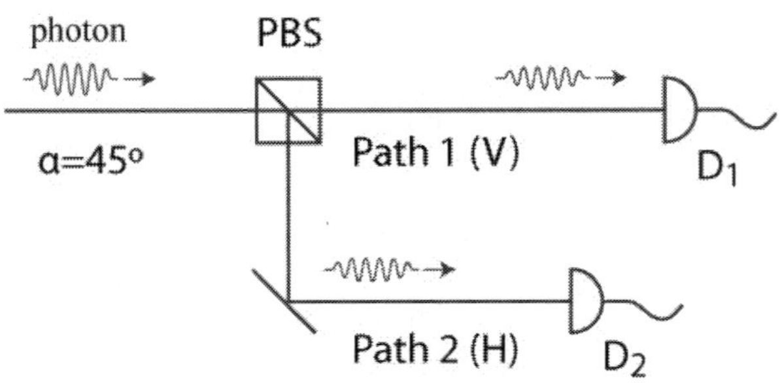

Figure 2.
Polarizing beam-splitter (PBS) experiment
with two detectors D_1 and D_2 in QM interpretations.

Some quantum optics physicists believe that photon does not make this decision *inside* the PBS. Each photon emerges from the PBS in the superposition of two equal possibilities: Path 1 with the vertical polarization (V) over Path 2 with the horizontal polarization (H).

Suarez entitled his paper as "Decision at the beam-splitter, or decision at detection, that is a question." His answer is, "photon makes decision at detection." [8]

According to Zeilinger, an individual photon decides which path to take only at the moment it is registered by one of two detectors (in our case, D_1 or D_2). "A photon actually does *not* decide at the moment when it leaves the beam-splitter what to do. ... The superposition collapses in the moment when the photon is registered by either one of the [D_1 or D_2] detectors.... And only at *that* moment does the photon decide which path it took." [9]

Such explanation is not satisfactory. In principle, both detectors can be placed deep in space and separated by several light years. Then instantaneous collapse of the superposition over any distance dramatically illustrates a QM conflict with special relativity.

This is Enigma 6.

As one can see, quantum mechanics leads us into quantum enigma quagmire. Photon physical indivisibility and scientific validity of special relativity would have to be put under question. It would be a high price to pay to maintain the myth that "quantum mechanics has never been proven wrong."

Even more important, by accepting the decision at detection would mean that *the present* influences *the past.*

If photon makes the decision at detection, then we are dealing with two events: event 1 - a single photon passing through the PBS, and event 2 - "click" at the detector. This means event 2 (the present) influences event 1 (the past). This is especially dramatic if both events are separated by several light years.

This is Enigma 7.

3.6. "No Click" Action by Photon Detector

In the layout shown in Figure 3, let us assume that we remove one detector, D_2, and keep in place the other detector, D_1. If detector D_1 clicks, then we know that the superposition collapsed, the photon selected Path 1 and was registered by D_1.

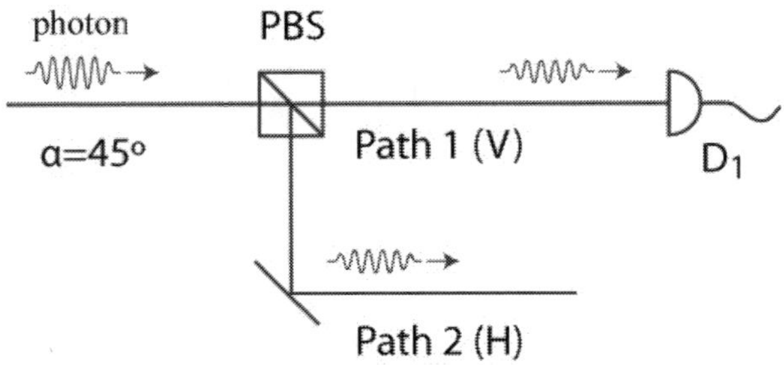

Figure 3.
Polarizing beam-splitter (PBS) experiment with single detector D_1 in QM interpretation.

However, if after sending a single photon to the PBS, (an event confirmed by a gating photon, to be explained later) a detector D_1 does not "click", then we know that the photon is traveling along Path 2. Somehow, the photon has made its decision without being registered by the only detector we have, namely D_1. In such case, even without registration by the detector, we know exactly which path the photon decided to take.

It implies that "*no click*" is sufficient to collapse the superposition and direct photon along Path 2.

This is Enigma 8.

The collapse of the superposition with "no click" cannot be explained within the existing paradigm of quantum mechanics. In any case, probabilistic quantum mechanics has difficulties in explaining individual quantum entities or processes. Quantum mechanics deals successfully only with assemblies. As Einstein said, a quantum theory which does not explain reality of individual quantum entities or processes is not a real thing.

In fact, the collapse of the superposition, caused by the "no click" detector action, is an earth shattering event. But it appears that no one has bothered to notice the fact that *non-physical action by detector D_1 causes the collapse of the superposition.* No explanation of this fact can be found within the existing QM paradigm.

This is Enigma 9.

3.7. Quantum Mechanics has no ontological explanation

The PBS experiment is the simplest experiment in quantum optics. Still, as one can see, the experiment results in a quantum enigma quagmire.

Here is a summary of some enigmas unresolved in QM:

How is it possible for the photon, with no physical dimension in the direction of travel, to interfere with itself?

How is it possible to explain the transformation of the circular polarization into the linear polarization for a single photon?

How is it possible for physically indivisible photon to travel along two paths at the same time separated by any distance?

How is instantaneous collapse of the superposition (reduction of wave function) over any distance possible, without getting into tension with special relativity?

How is it possible for the photon detector in "no click" action to cause the collapse of the superposition?

How is it possible that the present (click at detector) influences the past (the decision at the PBS)?

All these unresolved enigmas are a strong indication that the QM paradigm is fundamentally deficient. The QM paradigmatic deficiency is fully exposed because, in our analysis, we deal not with assemblies but with individual photons.

Something is missing in QM's understanding of fundamentals of quantum reality.

Chapter 4.

The Photon Inner Structure and Properties According to Aphysical Quantum Mechanics

4.1 The photon and the quantum reality: physical, aphysical, and elementary consciousness

QM recognizes only the physical quantum reality and is blind to the other fundamental categories. In QM, the photon operates only in the physical quantum reality.

In contrast, AQM dramatically expands the understanding of the quantum reality by introducing two additional fundamental categories: the aphysical quantum reality and the elementary consciousness of elementary particles.

In AQM the photon operates within all three fundamental categories of the quantum reality: the physical quantum reality, the aphysical quantum reality, and the elementary consciousness.

4.2 Photon three-dimensional inner structure: the physical energy c-ring, the aphysical energy cylinder, and the elementary consciousness.

AQM states that all elementary particles, including the photon, are not point-like, all have three dimensional inner structures. In fact, it is a travesty to insist that elementary particles are point-like and structureless. Figure 4 shows an example of the complete three-dimensional inner structure of the full-fledged circularly polarized photon.

♦ ♦ ♦

Postulate 31. *The inner structure of the photon (see Figure 4)*

The three-dimensional inner structure of the photon is the cylindrical system consisting of the physical energy c-ring and the aphysical energy hollow cylinder (AC), both with zero wall thickness. The c-ring and AC have the same Compton radius and the same Compton angular velocity. The c-ring and AC are both positioned along the same axis, with the c-ring encapsulated within AC. The physical energy c-ring has zero length and zero cross-section for boson such as the photon. As a result, the cross-section of photon-photon interaction is zero.

♦ ♦ ♦

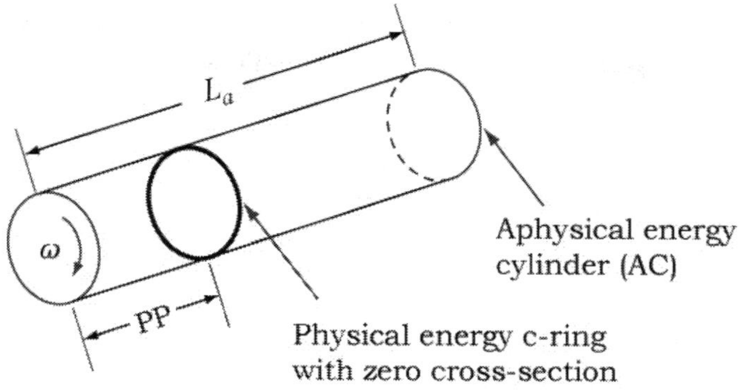

Figure 4.
The three-dimensional inner structure of the circularly polarized photon. PP – position parameter

4.3 Photon Position Parameter and its fundamental significance

The position of the c-ring relative to AC is defined as the position parameter, a unique characteristic of individual photon. As shown in Chapter 8, it is possible to determine experimentally the aphysical length of the photon, thus confirming the AQM photon model.

♦ ♦ ♦

Postulate 32. *Position Parameter as unique characteristic of individual photon.*

The position of the physical energy c-ring relative to the aphysical energy cylinder (AC) along their common spin axis is a unique characteristic for each individual photon. It is called the "Position Parameter" or PP (see Figure 4).

The Position Parameter is a source of fundamental irreducible quantum randomness. In a large assembly of monoenergetic photons, there are no two photons with exactly the same value of the Position Parameter. In any large photon assembly of monoenergetic photons, the distribution of the position of physical energy c-ring along the length of the aphysical energy cylinder is uniform (see Figure 5).

♦ ♦ ♦

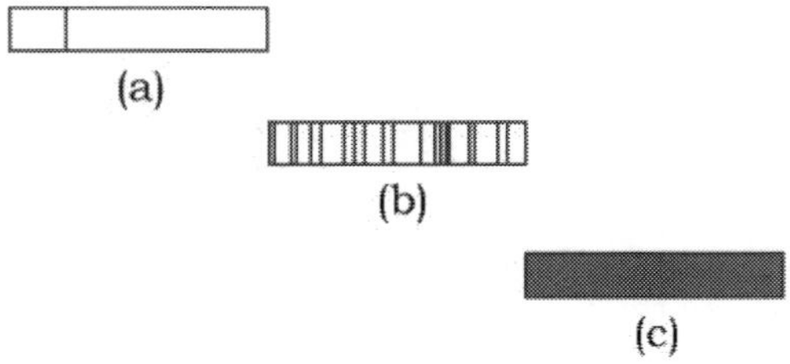

Figure 5.
The distribution of the position of the physical energy c-ring defined by the position parameter along the length of the aphysical energy cylinder.

(a) Individual photon, (b) Small assembly of photons.
(c) Large assembly of photons

4.4 Properties of the photon c-ring. Concept of the inversion and the inverted electromagnetic energy

The properties of the photon physical c-ring will be explained separately from the properties of the aphysical cylinder (AC) and performance of the elementary consciousness. To be consistent with special relativity, the photon physical energy c-ring has zero length. There is no mathematical complexity here. Furthermore, the photon c-ring, as any bosonic c-ring, has zero cross-section. All photon physical energy is confined within the photon c-ring with zero cross-section and, hence, in zero space.

The photon c-ring is the *bundle* of physical energy spinning with the circumferential speed of light at Compton angular velocity $\omega = c / r$, where r is Compton radius of the bundle. To satisfy special relativity, the bundle must have zero cross-section.

According to SM, the photon is called the quantum of electromagnetic energy (or quantum of electromagnetic fields) and is the force carrier for electromagnetism. Something is wrong with the SM photon definition. It is another SM fundamental misconception. *The properties of electromagnetic energy would not allow such energy to be totally "squeezed" into the c-ring or, in another words, to be totally confined into zero space.*

The photon is not the quantum of electromagnetic energy.

Here I introduce and explain the concept of the inversion and the inverted electromagnetic field (energy).

The fundamental nature of the inversion of the electromagnetic field (energy) into the inverted electromagnetic field (energy) confined in the c-ring with zero cross-section, is demonstrated in the following three examples.

Example 1. The annihilation of an electron-positron pair producing a pair of gamma-photons (γ_e):

$$e^- + e^+ \rightarrow \gamma_e + \gamma_e \qquad (4\text{-}1)$$

The annihilation of electron-positron pair, as shown above, is well known in particle physics. However, as explained in Volume 3, the process is much more complex than assumed in SM.

According to AQM, the electron and the positron are composite elementary particles. Both the electron and the positron have equal self-masses and therefore, equal radii.

The process of the annihilation can be presented in detailed spacetime dynamics. In AQM, the process is not called “the annihilation” but instead, the transformation of electron-positron pair into a pair of γ_e photons.

For the purpose of example 1, I utilize only a portion of the process, namely the physical process of the collision of intrinsic electron c-ring with intrinsic positron c-ring and the transformation into a pair of photon c-rings:

$$(\text{c-ring})_{\hat{e}^-} + (\text{c-ring})_{\hat{e}^+} \rightarrow (\text{c-ring})_{\gamma_e} + (\text{c-ring})_{\gamma_e}$$

Figure 6 shows two c-rings: $(\text{c-ring})_{\hat{e}^-}$ and $(\text{c-ring})_{\hat{e}^+}$ prior the collision, heading towards each other.

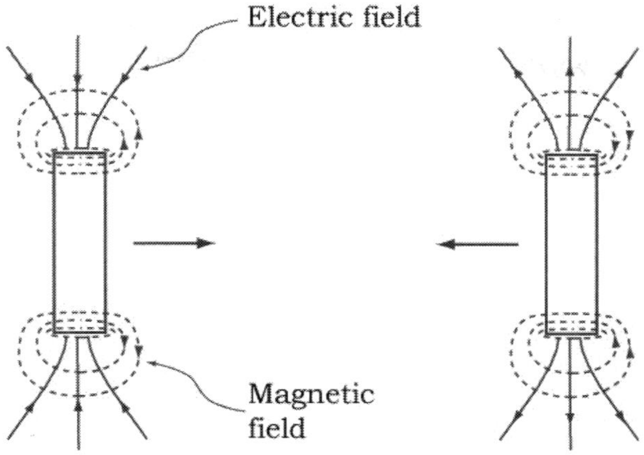

Figure 6.
The collision of intrinsic electron and intrinsic positron
(only physical energy c-rings are shown)

According to AQM, the intrinsic electron (or the intrinsic positron) is subset of the electron (or the positron). Each element in electrostatic and magnetostatic fields of the intrinsic electron c-ring has a corresponding anti-element in electrostatic and magnetostatic fields of the intrinsic positron c-ring (see Figure 6).

During the process of the transformation, intrinsic electron (and symmetrically, intrinsic positron) electrostatic and magnetostatic energies E_e and E_m are transformed into inverted electrostatic and inverted magnetostatic energies I_e and I_m of photon c-rings, with the conservation energy law observed (see Figure 7):

$$E_e = I_e \text{ and } E_m = I_m$$

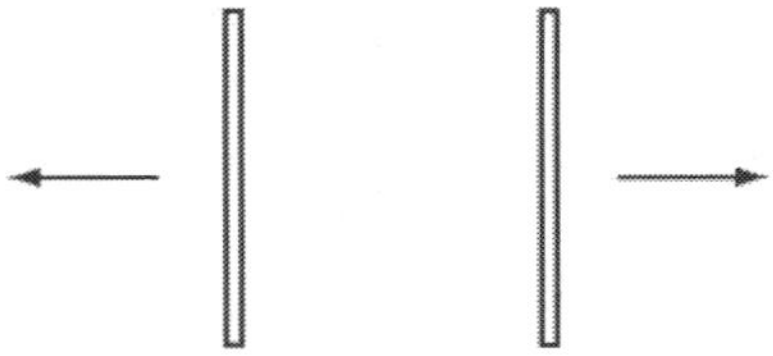

Figure 7.
The production of pair of γ_e photons from the annihilation of electron-positron pair
(only photon c-rings are shown)

♦ ♦ ♦

Postulate 33. *The transformation of the Electromagnetic Field into the Inverted Electromagnetic Field in Zero Space of the Photon C-ring.*

A collision of intrinsic electron-positron pair of equal kinetic energy results in the transformation of their electrostatic and magnetostatic fields of unlimited space into inverted electrostatic and magnetostatic fields in zero space of photon c-rings.

The physical energy of the intrinsic electron-positron pair distributed in unlimited space is transformed into the inverted physical energy in zero space of photon c-rings.

In contrast to SM, the photon is the quantum of the inverted electromagnetic energy.

♦ ♦ ♦

♦ ♦ ♦

Postulate 34. *AQM Absolutes.*

AQM includes several absolutes, such as:

- *the photon physical energy c-ring with zero length, zero radial thickness, and zero cross-section that encloses inverted electromagnetic energy;*
- *the photon aphysical energy cylinder with zero radial thickness;*
- *the electron physical energy c-ring with zero radial thickness;*
- *instantaneous reconstruction of the self-entangled quantum state into the full-fledged quantum state, regardless of distance separating a-fractions from the host;*
- *instantaneously transmitted influence from one entangled quantum entity to another one, regardless of separating distance.*

♦ ♦ ♦

Further explanation of the absolutes must be left to scientists in the distant future. Any attempts to explain the absolutes would lead to endless and mindless speculations. Our science is only six hundred years old and still too young for such a challenge.

Example 2. Synchrotron radiation.

Synchrotron radiation is produced by ultra-relativistic charged particle, such as electron, crossing strong static magnetic field, as shown in Figure 8.

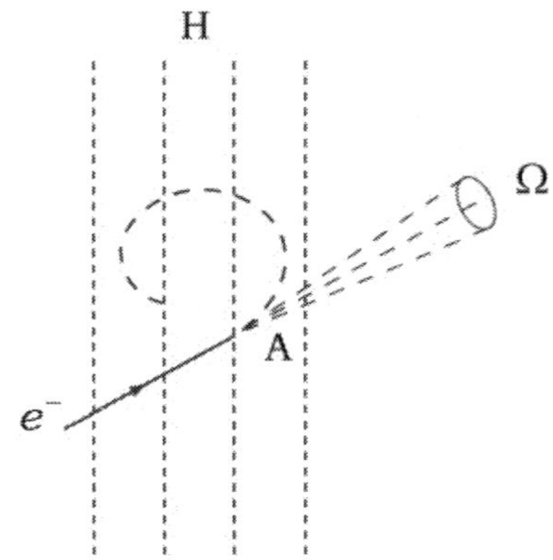

Figure 8.
An ultra-relativistic electron crossing static strong uniform magnetic field H.

Total power radiated by electron within sharply collimated solid angle Ω is proportional to $\frac{1}{m^2}$, γ^2 and H^2, where $\gamma = \frac{1}{\sqrt{1-\beta^2}}$ is the relativistic factor.

The AQM explanation of the mechanism for the synchrotron radiation is presented in Postulate 35.

♦ ♦ ♦

Postulate 35. *Mechanism of Synchrotron Radiation.*

When ultra-relativistic electron crosses strong magnetostatic field, electron is subjected to enormous stress by two conflicting forces:

Force 1 is trying to keep electron magnetic moment (and spin) aligned along the direction of the magnetic field;

Force 2 is trying to keep magnetic moment (and spin) aligned along the direction of electron travel.

Under such stress, electron converts some of its electromagnetostatic energy into inverted electromagnetic energy in the form of a collimated stream of photons.

♦ ♦ ♦

Comment: I offer a general concept of an extreme synchrotron radiation experiment for the future International Linear Collider with expected output electron energy of 125 GeV and short pulsed (a few microseconds) magnetic field with a strength in Mega-gauss range generated in small space.

Example 3. Elastic scattering of two electrons with equal kinetic energies, shown in Figure 9.

At close distance, both electrons are subjected to stress resulting in the conversion of the electromagnetostatic energies into photon inverted electromagnetic energies via a mechanism of symmetrically emitting mediating photons in the direction of their momenta causing repelling force additionally to Coulomb force, thus creating impression that electric charge is larger than it actually is, with the energy conservation law observed. The elastic electron-electron scattering produces equal change in each electron's individual momentum with some change in individual kinetic energies. *Mediating photons are real photons.* Their detection is difficult but not impossible.

The vacuum polarization is another SM fundamental misconception. The vacuum polarization does not exist. *"The charge screening" is a result of photon mediating action.* Again, no vacuum polarization is needed for explanation of "charge screening".

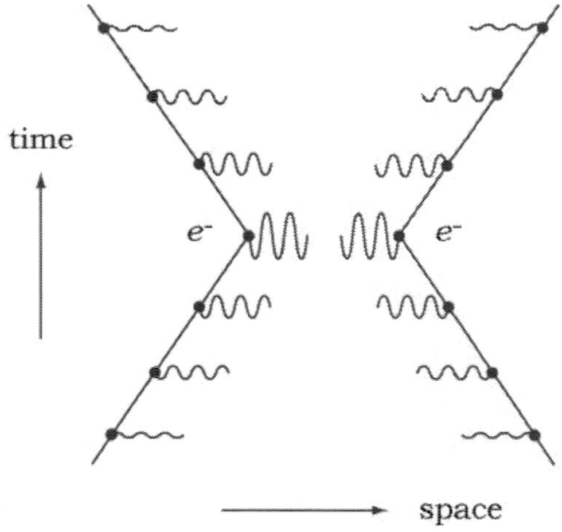

Figure 9.
Elastic interaction of two colliding electrons resulting in emission of mediating photons.

The subject of the charge screening and the vacuum polarization is expanded in Volume Three.

4.5 The photon c-ring as anisotropic inverted electromagnetic energy bundle

The photon is an anisotropic boson. The photon's inverted electromagnetic energy I_{em} is distributed circumferentially within the photon c-ring as an energy bundle with amplitude:

$$a_{em} = a_e \cos \omega t + a_m \sin \omega t \, ,$$

see Figure 10 (a), and density:

$$d_{em} = d_e \cos^2 \omega t + d_m \sin^2 \omega t \, ,$$

see Figure 10 (b), where $d_e = I_e/\lambda$, $d_e = I_m/\lambda$, $I_e = I_m$, $\lambda = 2\pi r$, and $I_{em} = I_e + I_m = \hbar\omega$.

Photon c-ring circumferential inverted electromagnetic density, $d_{em} = d_e + d_m$, is constant,

$$d_{em} = hc/\lambda^2 \, .$$

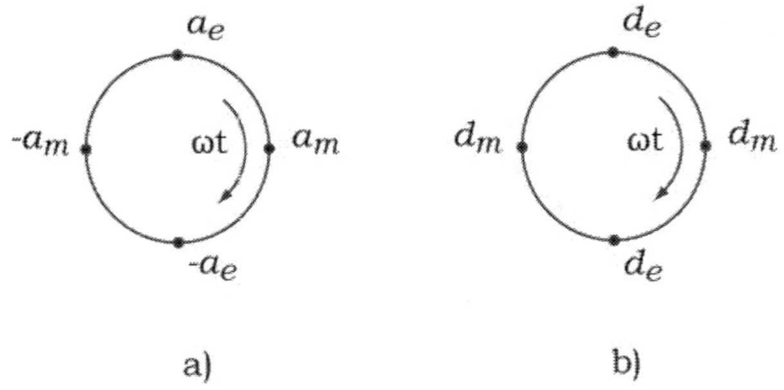

Figure 10 (a), (b).
Inverted amplitude (a) and inverted energy density distribution (b) along the photon c-ring circumference

4.6 Zero cross-section of interaction in photon-photon collision

There have been several attempts to study photon-photon interactions experimentally using colliding beams of high power lasers. An example is presented in [10]. Crossing high-intensity laser beams are expected to produce *signal photons* with frequency and direction propagation different from the direction of laser photons. But the results have been inconclusive.

Such experiments are then followed up with a discussion about vacuum emission, fluctuations of virtual charged particles, and quantum vacuum nonlinearities.

AQM view denies such vacuum emission, fluctuations of virtual charged particles, and quantum vacuum nonlinearities.

According to AQM, the photon c-ring, with the inverted electromagnetic energy bundle, has non-zero radius $r = c/\omega$ but *zero cross-section.*

As an example, let us consider the collision of two photons with different energies. Such a process can be presented as two c-rings with zero cross-section crossing each other, resulting in no photon-photon interaction, shown in Figure 11.

The universe is filled with photons traveling isotropically in all possible directions, i.e. the cosmic background radiation. Photons are crossing each other with no interaction. This is the principal reason why photons can travel for billions of years reducing gradually their energy due to space expansion. Photons are not point-like as stated in SM. Photon size is defined by its radius $r = \lambda/2\pi$. With the photon c-ring having zero cross-section, no photon-photon collisions are expected.

To prove that signal photons can be experimentally produced in extreme high-intensity laser experiments would be like proving that a value of the speed of light in free space has finite although extremely small distribution. That is a high order task which should be left to future generations of scientists.

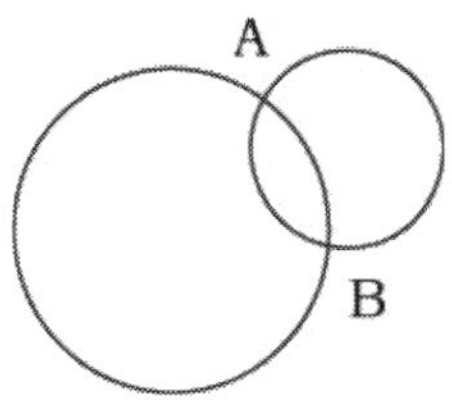

Figure 11.
An example of collision of two photon c-rings of different energies resulting in no interaction.

4.7 Photon polarization quantum states: circular, linear, and elliptical

A) Photon circular polarization.

One can present circularly polarized photon spin S in simple terms as:

$$S = mvr ,$$

where $m = \hbar\omega / c^2$ is photon self-mass in its system of self-reference; $v = \pm c$ is c-ring circumferential velocity; and $r = c/\omega$ is c-ring Compton radius.

The circularly polarized photon has spin $S=\pm1$ in units of $\hbar$.

$S=+1$ is for a right handed photon with spin in the direction of travel, shown in Figure 12 (b).

$S=-1$ is for a left handed photon with spin opposite to the direction of travel, shown in Figure 12 (a).

B) Photon linear polarization, shown in Figure 12 (c).

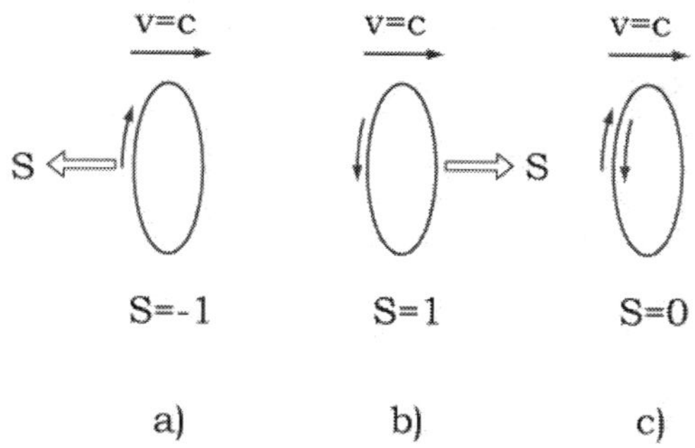

Figure 12 (a), (b), (c).
Photon three quantum spin states *S* and their respective bundle of energy configurations: a) left-handed; b) right-handed; c) linearly polarized

In the case of the linear polarization, the bundle of photon physical energy in the c-ring is split into two equal sub-bundles. Sub-bundles rotate opposite to each other with the Compton angular frequency ω at the Compton radius $r = c/\omega$ within c-ring zero space.

In the visible part of the spectrum, the photon radius is in the range of 70 nm to 110 nm (1 nm = 10^{-9} m).

Each bundle has sub-spin S_s defined as

$$S_s = mvr = \frac{E_S}{c^2} c \frac{c}{\omega} = \pm \frac{\hbar}{2},$$

where $E_S = \frac{1}{2} E = \frac{1}{2} \hbar\omega$ is the energy of each sub-bundle.

As shown in Figures 10 (a, b), each bundle of physical energy within the c-ring is a source of inverted electrostatic a_e and inverted magnetostatic a_m fields distributed along its circumference described as

$$a_e = a_e \max \cos \omega t$$

and

$$a_m = a_m \max \sin \omega t \, .$$

Inverted electromagnetostatic energy density along the c-ring circumference is constant.

As shown in Figures 12 (a, b, c), the bundle(s) are configured in three quantum spin polarization states:

1. Single energy bundle with circular right-handed polarization aligned in the direction of photon momentum (S=1);
2. Single energy bundle with circular left-handed polarization aligned opposite to the direction of photon momentum (S=-1);
3. Two equal sub-bundles producing linear polarization with spin S=0.

In the case of the linear polarization, an important feature of the photon inner structure is that both equal energy sub-bundles share the same zero space. In free space, two energy sub-bundles cannot be separated.

In the case of quantum spin state S=0, the plane of the linear polarization is defined by angle θ, responding to crossing point of maxima of inverted electric fields of both bundles, as shown in Figure 13. Angle θ and the plane of the linear polarization can be experimentally adjusted.

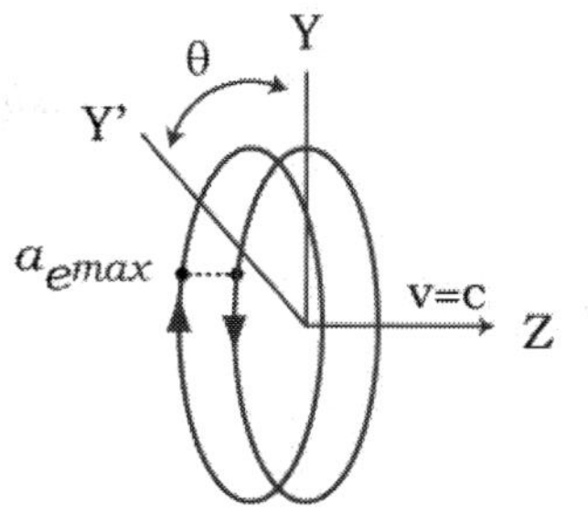

Figure 13.
Photon plane of the linear polarization Y'Z at angle θ, with respect to the reference plane YZ.

C) Photon elliptical polarization.

In quantum optics research laboratories, it is routinely demonstrated that individual photons can be brought to the elliptical polarization state by using standard optical components. By using the AQM model of photon, one can explain the elliptical configuration as an adjustable ratio of the left-handed sub-bundle with energy E_L and the right-handed sub-bundle with energy E_R, while total photon energy $E = E_L + E_R$ remains constant.

Photon spin can be presented as $S = S_L + S_R$, where $S_L = \hbar E_L / E$ and $S_R = \hbar E_R / E$. In the three cases discussed above, we have

$S = 1$ for $E = E_L$ and $0 = E_R$,

$S = 0$ for $E_L = E_R = 1/2 E$,

$S = \text{-}1$ for $E = E_R$ and $0 = E_L$.

In the range $E_L/E=1$ to $E_L/E=0$ (or from $E_R/E=0$ to $E_R/E=1$), the photon spin varies in a range of [1 to -1], thus showing its incredible versatility, at least in the optical energy range.

In conclusion, photon elliptical configuration can be presented in two equally valid interpretations:

- as an unequal split of photon c-ring energy bundle into opposite rotating circular polarization sub-bundles (Figure 14 (a)); or
- as an unequal split of photon c-ring energy bundle into the circular polarization and the linear polarization sub-bundles (Figure 14 (b)).

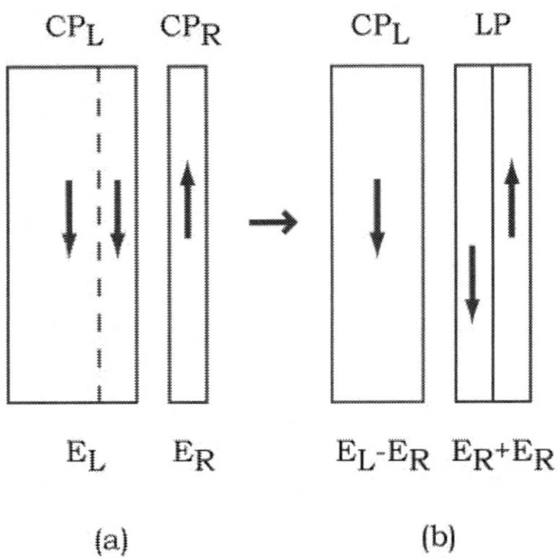

Figure 14 (a), (b). Photon elliptical polarization.

The elliptical polarization is superposition of two quantum states (a) circular and circular; (b) circular and linear.

♦ ♦ ♦

Postulate 36. *Ontology of Photon Polarization.*

Photon right-handed (R-H) circular polarization is produced by right-handed spin (R-H rotation) of the entire photon c-ring physical energy bundle.

Photon left-handed (L-H) circular polarization is produced by left-handed spin (L-H rotation) of the entire photon c-ring physical energy bundle.

Photon linear polarization is produced by one half of the photon c-ring physical energy bundle spinning in the R-H rotation and another half spinning in the L-H rotation.

Photon elliptical polarization is produced by an unequal portion of photon c-ring physical energy bundle spinning in R-H rotation and the remainder spinning in L-H rotation.

♦ ♦ ♦

4.8 Photon-photon entanglement

AQM provides ontological explanation for photon-photon entanglement.

♦ ♦ ♦

Postulate 37. *The Photon Entanglement.*

Two photons traveling in free space are entangled if elementary consciousness residing in a c-ring of one photon is connect-

ed via c-link with elementary consciousness residing in a c-ring of another photon, as shown in Figure 15.

♦ ♦ ♦

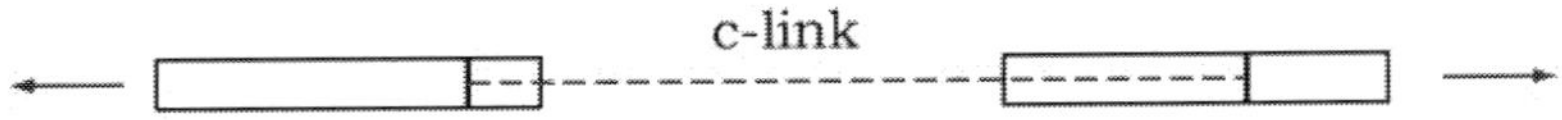

Figure 15.
Photon-photon entanglement.

For the most part, the universe is transparent to photons due to photon-photon zero cross-section of interaction. That is also valid for entangled photons as well.

Two entangled photons can travel in the universe for billions of years with no disruption to their entanglement. Two entangled photons are connected by their c-link and no cosmic object can cut or destroy c-link. Photons, separated by billions of light years, are still connected by their c-link. C-link crosses stars, galaxies, even black holes and survives (see Figure 16).

In contrast, based on properties of Bose-Einstein condensate and experimental evidence, two entangled bosonic atoms (such as H_2 or N_2), once exposed to free space, could not survive even a fraction of a second. Their entanglement would be immediately disrupted by the cosmic background radiation.

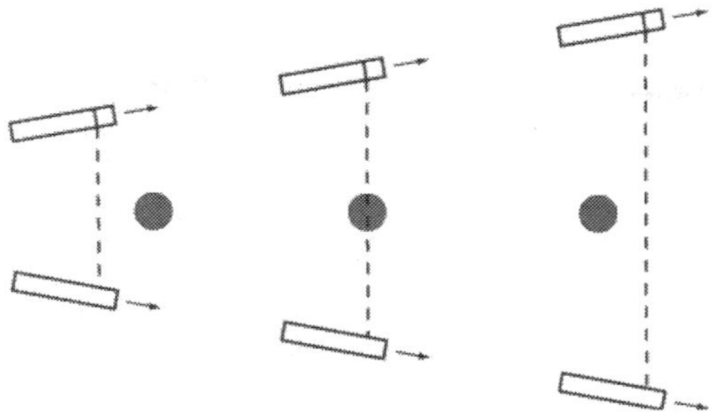

Figure 16.
The c-link of two entangled photons crossing a cosmic object in the vast space expanse of the universe.
Notation: ● – a cosmic object (e.g., neutron star).

4.9 Photon self-entanglement and spacetime dynamics of the reconstruction into the full-fledged quantum state

Self-entanglement is a massive property of all elementary particles. The subject of self-entanglement is discussed in great detail in Vol. 1. Here we address the subject specific to optical photons in the visible part of the spectrum with wavelength in the range of 0.4 – 0.7 μm.

♦ ♦ ♦

Postulate 38. *Aphysical-aphysical interaction of optical photons in the optical domain.*

The interaction of optical photons with ideal optical material is not a physical process. Such process is governed exclusively by aphysical energy via aphysical-aphysical interaction. In a "no physical energy transfer environment", such as transparent optical materials in quantum optics, it is the aphysical substance of the photon, as a set of a-fractions including the host, that plays the role in optics dynamics. The aphysical substance is the deterministic player in quantum optics. That is why quantum optics should be called 'aphysical quantum optics.'

♦ ♦ ♦

Photon self-entanglement means that the photon aphysical cylinder is split into several cylindrical fractions, called *a-fractions*, where only one of them carries the physical substance, the c-ring. Such special a-fraction is called the *host*. The self-entanglement is shown in Figure 17.

The c-ring by itself does not have self-entanglement. The c-ring is always located intact in one of a-fractions, the host. Each a-fraction in its shape and form is identical to the aphysical energy cylinder of the full-fledged photon prior to its self-entanglement but with reduced aphysical energy density. Angular velocity is the same for all individual a-fractions and is equal to Compton angular velocity of the full-fledged photon prior to its self-entanglement. Each individual a-fraction has the image of c-ring ("cr-image"), located precisely in the same position as the c-ring (see Figure 18). Values of position parameter are equal for all a-fractions including the host.

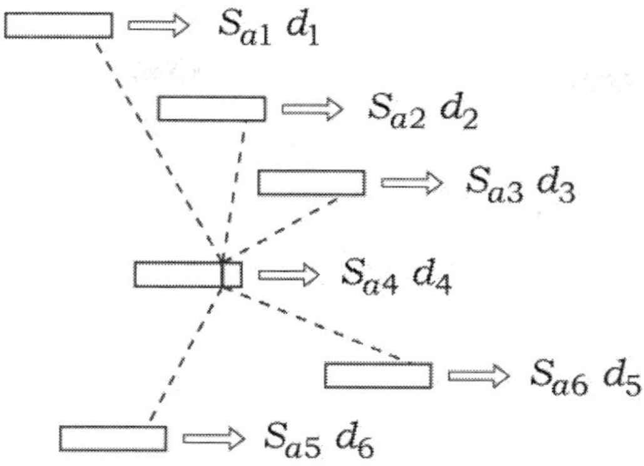

$$S_a = \sum S_{an}\ ,\ d = \sum d_n = 1$$

Figure 17.
The photon self-entanglement.
Aphysical energy densities d_n and aphysical fractional spins S_{an} are shown for individual a-fractions, including the host. As a general rule, directions of travel of individual a-fractions are not aligned with the host.

What distinguishes one a-fraction from another one is the amount of aphysical energy, called aphysical density d_n, it carries and a value of aphysical spin S_{an}, where d_n is the normalized aphysical energy density of a-fraction n, including the host, with a total normalized aphysical energy density

$$d = \sum d_n = 1.$$

An individual value of aphysical spin of a-fraction n is defined as

$$S_{an} = d_n\, \hbar/U$$

where U is the universal constant. The aphysical spin of a fully-fledged photon S_a is

$$S_a = \sum S_{an} = \hbar/U$$

The host c-ring is connected to each individual a-fraction cr-image via c-link, from c-ring to cr-image, as shown in Figure 18.

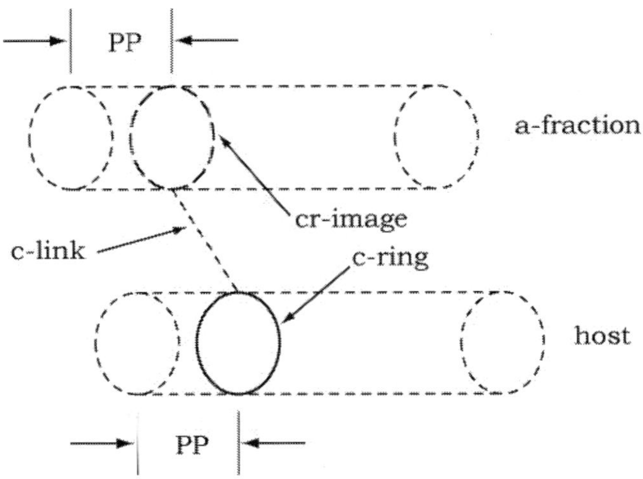

Figure 18.
The host and a-fraction are connected by c-link from c-ring to cr-image.

Elementary consciousness, which resides in the host c-ring, is aware of the status of each individual a-fraction and is in control. This sounds incredible to physicists with a materialist mindset.

To a significant extent, aphysical properties of the photon imitate its physical properties. For example, the inverted aphysical electromagnetic field imitates the inverted physical electromagnetic field. The aphysical spin is an imitation of the physical spin.

♦ ♦ ♦

Postulate 39. *Aphysical-aphysical interaction in interference and diffraction.*

Within optical material or in free space, an individual a-fraction behaves as "a full-fledged photon" until it attempts aphysical-physical interaction with another quantum entity ("foreigner"). Such interaction is forbidden and a-fraction is instantaneously recalled by the host elementary consciousness. The process is defined as the partial reconstruction. However, an individual a-fraction is allowed to participate in any aphysical-aphysical interactions with any aphysical entity, which typically results in a change of its travel direction. Aphysical-aphysical interaction is always elastic with no transfer of aphysical energy. The aphysical-aphysical interaction plays the principal role in interference and diffraction.

♦ ♦ ♦

Once a-fraction is recalled, it joins the host and aligns its spin with the host's spin. In the case of total reconstruction, total aphysical spin S_a is reconstructed.

4.10 Photon splitting into secondary photons

Photon is the most versatile elementary particle. In addition to the many properties discussed previously, there is one more: photon splitting into secondary photons, observing the conservation laws and timing.

The splitting of a primary linearly polarized photon into two secondary linearly polarized photons is accomplished by using a non-linear crystal, a technique known in quantum optics as "spontaneous parametric down conversion" or SPDC [11, 12]. SPDC is a single-photon source of entangled linearly polarized photon pairs, with one photon called a signal photon and another a gating photon. The source of primary photons is a laser, or "pump". The process is shown in Figure 19, with conservation laws described:

$$\hbar\omega_p = \hbar\omega_s + \hbar\omega_g ,$$

$$\omega_p = \omega_s + \omega_g ,$$

$$\hbar\bar{k}_p = \hbar\bar{k}_s + \hbar\bar{k}_g ,$$

$$\bar{k}_p = \bar{k}_s + \bar{k}_g .$$

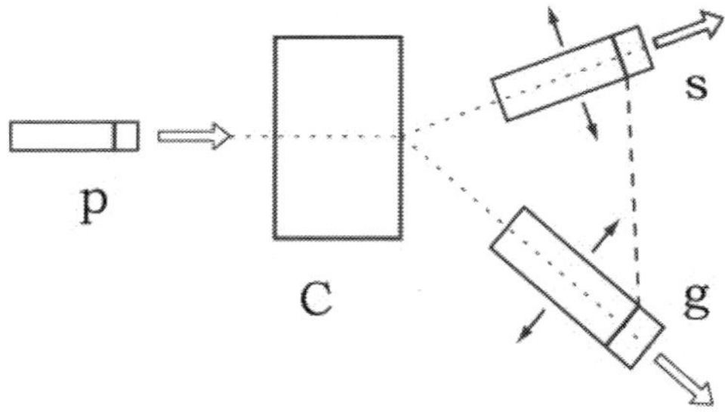

Figure 19.
Spontaneous parametric down conversion.
p is the primary photon, linearly polarized in horizontal plane;
C is the non-linear down-conversion crystal;
s is the signal photon, linearly polarized in vertical plane;
g is the gating photon, linearly polarized in vertical plane.

SPDC is a valuable experimental technique as a single-photon source (one photon at-a-time) of pairs of correlated or entangled photons. When a gating photon is detected, a signal photon is expected to arrive at the signal detector. All other "spurious" photons are filtered out from recording. In Chapter 8, I show how to use the SPDC for measurement of *photon aphysical length. After the normalization by wavelength, the photon aphysical length is the AQM fundamental aphysical constant.* The experiment is based on AQM interpretation.

In principle, nothing can prevent secondary photons from splitting further.

4.11 Photon stability and challenge to special relativity

The photon is stable. Unless it is physically obstructed, a photon travels for billions of years and covers an immense expanse of the universe, gradually reducing its energy due to space expansion. It does not travel along a straight line but as prescribed by spacetime geometry, in accordance with general relativity proving that the photon has self-mass.

The AQM photon model shows peculiar, possibly unique, properties of special relativity. The inner structure of the photon involves two orthogonal motions, both at the speed of light: one longitudinal in the direction of photon travel, and second one rotational in the transverse plane to the direction of travel. Both motions are strictly orthogonal with respect to each other. The orthogonal requirement is "absolute."

This photon property represents a challenge to special relativity.

♦ ♦ ♦

Postulate 40. *The spinning c-ring as the origin of the photon self-mass.*

Photon self-mass E/c^2 is generated by the spinning c-ring. The origin of the photon self-mass is the c-ring spinning energy. Surprisingly, the photon has no intrinsic kinetic energy. The photon spinning c-ring self-mass is the same in either its own frame of self-reference $v = c$ or any other frames of reference. The photon spinning c-ring self-mass is invariant. As photon interacts with other absorbing physical entities, photon spinning c-ring self-mass is transformed into kinetic energy just before photon is destroyed and absorbed.

♦ ♦ ♦

Longitudinal motion cannot generate kinetic energy for a simple reason: the photon has zero length longitudinally in the direction of travel, therefore it has no length to contract. Without length contraction, there can be no kinetic energy. This is a counterintuitive point which is difficult to grasp.

One can argue that a reason the photon has zero longitudinal length is because it travels with the speed of light. Such reasoning is faulty however – it leads to infinite kinetic energy.

At first glance, it appears that for a photon to be stable during its travel in free space, it needs the centripetal force to balance its centrifugal force F:

$$F = E / r .$$

Photon energy is inversely proportional to its radius:

$$E = \hbar c / r$$

The centrifugal force is inversely proportional to square of its radius:

$$F = \hbar c / r^2 .$$

The centrifugal force exists but the centripetal force is missing.

It has not yet been discovered. Some of you, readers, will say, "This is incredible! It cannot be!". I agree with you but one cannot argue with objective reality, whether one likes or not. I offer you an alternative explanation in the form of a postulate.

♦ ♦ ♦

Postulate 41. *No pathway exists for photon traveling in free space to spontaneously convert itself into another inner structure.*

Traveling in free space, a photon cannot release its energy by expanding. AQM denies physical interaction between elementary particles and the intrinsic vacuum. Speaking in a simple layman language, the photon cannot release energy in the vacuum because there are no "receiving hands". There is no pathway for a photon traveling in free space to convert itself into another inner structure.

♦ ♦ ♦

However, once a photon collides with another elementary particle, such as electron, neutrino, or quark, the photon instantaneously self-destructs and releases its total energy or sometimes converts into a secondary photon by a process known as Compton scattering.

Photon reduces its energy and radius due to space expansion according to:

$$dE = -\left(\hbar c / r^2\right) dr .$$

The reverse is also true if a photon is falling into a gravitational field (e.g., a black hole). Here the photon increases its energy due to space contraction.

4.12. Photon spacetime dynamics. Tension with special relativity

So far, special relativity has been an uncontroversial theory in comparison with quantum mechanics, which has so many enigmas, numerous interpretations, and deficient paradigm.

Based on AQM, photon dynamics and its interactions represent a challenge to special relativity. In free space, a photon travels with the speed of light. Surprisingly, in the photon *transverse* plane there are none of the effects of special relativity. The physical energy bundle(s) rotates at the Compton angular velocity. Circumferential linear velocity is equal to the speed of light. In the photon transverse plane, there is no time dilation or space contraction. Photon wavelength remains unchanged and still measured as equal to the c-ring circumference. The period of c-ring rotation is *the period* with no effects of time dilation.

By now the reader knows that radial thickness of the c-ring is zero. Is this the effect of relativistic contraction? The answer is no. C-ring zero thickness is the result of the general Compton conditions which require the c-ring circumferential linear velocity to be equal to the speed of light.

Dynamics and the length of aphysical energy cylinder (AC) are unchanged regardless of the frame of reference. AC experiences no relativistic effects. The aphysical energy cylinder length is non-zero. If normalized by wavelength (to be more precise, by the AC circumference), it is an AQM fundamental aphysical constant L_a/λ, in any frame of reference, including the photon's own frame of self-reference.

Is special relativity heading toward a crisis? The answer—very likely.

4.13. Photon: Types of Interactions

Here is another aspect for our discussion.

Physical energy E of a relativistic particle with momentum p and self-mass m is defined by the established relation:

$$E^2 = m^2c^4 + p^2c^2 \qquad (4\text{-}1)$$

(1) *(2)*

Surprisingly, the relation (4-1) is not straightforward in its application to the photon. One has to specify the type of photon interaction, such as:

aphysical-aphysical;

aphysical-physical; and

physical-physical.

Case 1. Aphysical-aphysical interaction.

The aphysical domain is predominant in the photon optical energy range. It is defined as no-loss optical material, no-loss optical components, no-loss periodic structures, and no-loss (perfect) crystals. Here part (1) of relation (4-1) is applicable. Photon physical energy $E = mc^2 = \hbar\omega$; self-mass of spinning c-ring $m = \hbar\omega / c^2$; and spin $S = \pm\hbar$ at velocity $\upsilon = c$ and Compton radius $r = c/\omega$. Aphysical energy is equal to physical energy divided by U.

In such aphysical domain, aphysical energy is the driving force. Photon acquires a self-entanglement quantum state with no limit on the number of a-fractions. A-fractions can proceed to a secondary self-entanglement. There is no limit on the number of higher order of self-entanglement. Each a-fraction is connected by c-link to the host. Dynamics of a-fractions, including the host, is governed by the aphysical domain and the aphysical energy. Surprisingly, physical energy plays no role. This is the domain of *aphysical quantum optics.*

Case 2. Aphysical-physical interaction.

This, in fact, is not even an interaction – it is *a trigger.* Each individual a-fraction is allowed to interact aphysically with aphysical entities in the aphysical domain with no restriction. Such interactions are elastic. On the other hand, if an a-fraction meets a physical entity, which is not part of the aphysical domain, for example an impurity in the crystal structure, then the a-fraction is instantaneously recalled by the elementary consciousness residing in the c-ring. The process is called *partial reconstruction* of the full-fledged quantum state.

If the host meets a physical entity, such as impurity, then all a-fractions are recalled instantaneously ("the collapse of the wave function"). Photon is reconstructed from the self-entangled state to the full-fledged state. Only after that, the photon proceeds to physical-physical interaction with the physical entity and to energy transfer.

Case 3. Physical-physical interaction.

Let us consider a self-entangled photon leaving the aphysical domain and heading toward a photon detector. Just prior to interaction with the detector, photon is reconstructed into the full-fledged state, then at first touch with a detector, it undergoes further transformation and transfers both physical and aphysical spins to the detector. The process is governed by part (2) of relation (4-1). The process does not allow entanglement between photon and detector. It transforms its spin physical energy into physical kinetic energy.

During the process of physical-physical interaction, aphysical spin energy cylinder is transformed into aphysical kinetic energy column with length defined by de Broglie relation:

$$L_{ka} = h/p = \lambda$$

Photon physical and aphysical spins are transferred to detector.

The end result is photon being absorbed by detector and transformed into something else.

Measurements of photon aphysical-aphysical interactions are not available to science at this stage.

According to positivism, what cannot be measured or observed does not exist. I want to remind the reader of Einstein's quotation: "I am not a positivist. Positivism states that what cannot be observed does not exist. This conception is scientifically indefensible, for it is impossible to make valid affirmations of what people 'can' or 'cannot' observe. One would have to say that 'only what we observe exists' which is obviously false." [7]

I also want to remind the reader that we are an *emerging embryonic intelligence* [13]. Our science is only 700 years old. We have to recognize the fact that at this stage in our evolution, most of the objective reality is still not discovered and obviously not measured or observed. Objective reality exists regardless of our opinion.

The dynamics and interactions of the photon are so subtle and fundamental that it is worthwhile to summarize the process here in a postulate:

♦ ♦ ♦

Postulate 42. *Detailed energy transformation process of photon interaction in optical domain including detector.*

In the aphysical-aphysical optical domain, the photon is self-entangled. The host and a-fractions have independent trajectories. There is no role for physical energy. Aphysical-aphysical interactions are elastic and no exchange or transfer of aphysical energy takes place.

A photon heading toward a photon detector experiences two subsequent transformations. During aphysical-physical interac-

tion phase, the photon recalls a-fractions and becomes fully-fledged. The process is instantaneous.

In the next and final phase, the photon converts its physical spin energy and aphysical spin energy into photon kinetic energies. The process is governed by part 2 of relation (4-1): $p = E/c$. *Aphysical spin energy cylinder is transformed into aphysical kinetic energy column before it is transferred to detector. Both, physical and aphysical spins are transferred to the detector.*

♦ ♦ ♦

Chapter 5

A Polarizing Beam-Splitter Experiment with Linearly Polarized Photons Based on Aphysical Quantum Mechanics. The Resolution of Quantum Mechanics Enigmas

Here is a direct comparison of a conceptual quantum optics experiment in an AQM interpretation with the QM experiment described previously in Chapter 3.

A simple quantum optics experiment, using the PBS with linearly polarized photons and a single photon source, demonstrates incredible paradigmatic power of AQM over QM and resolves QM enigmas arising from the QM experiment. The simplicity of the experiment has no need for wave function or any mathematical formalism, and shows detailed visualized spacetime dynamics of a single photon.

All nine QM enigmas are resolved and ontologically explained.

The down conversion is used as a single-photon source, which delivers one photon at-a-time [14, 15]. The arrival of each signal photon is monitored by its twin photon (a gating photon) at a gating detector.

A single photon is a natural environment for AQM and an unnatural one for the probabilistic QM. The probabilistic nature of QM is one of the principal sources of QM enigmas.

As a probabilistic theory QM should not pretend on the foundational position.

AQM theory of the photon is presented in detail in Chapter 4. Here is summary of AQM principal and pertinent properties of a single photon.

Let us begin with the structure and properties of the circularly polarized photon as shown in Figure 4, Chapter 3.

A circularly polarized photon structure consists of:

1. the hollow aphysical energy cylinder (AC) with zero wall thickness, Compton radius $r = c/\omega$, and aphysical length L_a;

2. the physical energy c-ring with zero cross-section and the same Compton radius r embedded into the aphysical cylinder.

3. the c-ring which carries elementary consciousness.

The photon c-ring is a section of a hollow cylinder with Compton radius $r = c/\omega$, zero length and zero thickness, that means zero cross-section.

Photon c-ring is a spinning bundle of physical energy $E_p = \hbar\omega$, and photon wave length $\lambda = 2\pi r$. The circumferential velocity of c-ring physical energy bundle is the speed of light. The cross-section of physical energy bundle is zero. The photon c-ring physical energy bundle(s) is inverted electromagnetic energy.

Photon aphysical energy is $E_a = E_p / U$. The aphysical energy cylinder is an integral part of the photon inner structure. The aphysical cylinder and the c-ring are spinning with the same Compton angular velocity ω. *Aphysical energy cylinder contains inverted aphysical electromagnetic energy.* The length of aphysical cylinder is a multiple of the photon wave length representing an AQM fundamental aphysical constant L_a/λ. As explained in Chapter 8, the length L_a can be determined experimentally in quantum optics research labs or even in college undergraduate labs.

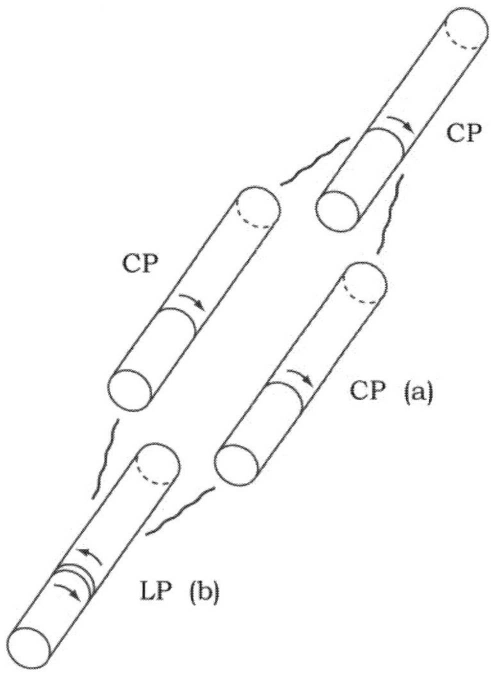

Figure 20.
Conversion of the photon circular polarization (CP) into the linear polarization (LP) and vice versa

CP(a) is converted into opposite helicity prior to producing LP(b)

Photon linear polarization is a superposition of two identical photon cylindrical substructures, spinning with equal but opposite Compton angular velocities. This resolves QM Enigma 3 (see Chapter 3) on how to produce the linear polarization from the circular polarization for a single photon.

In the case of the down-conversion, each primary photon with horizontal linear polarization is converted by a non-linear crystal into two correlated or entangled photons with the vertical linear polarization and physical energies $E_{p1} = \hbar\omega_1$ and $E_{p2} = \hbar\omega_2$, where $\omega_1 + \omega_2 = \omega$, $r_1 = c/\omega_1$ and $r_2 = c/\omega_2$.

In the down-conversion, one photon of a pair is used as a "signal" photon and the second one as a "gating" photon. When a gating photon is detected at gating detector, one is assured that a signal photon is delivered to signal detector. The down-conversion experiment layout is not shown here. It is further elaborated in Chapters 8 and 9.

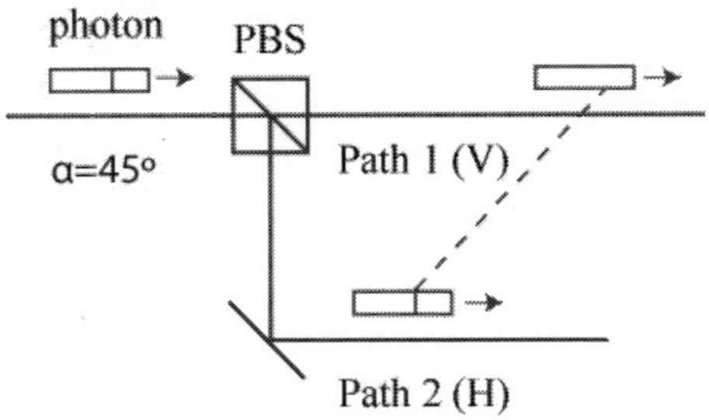

Figure 21.
Polarizing beam-splitter (PBS) experiment with linearly polarized photons; (V) – vertical polarization; (H) – horizontal polarization.

Figure 21 shows the PBS experimental setup. A single linearly polarized photon (primary input photon) is directed through PBS at angle $\alpha = 45°$, where α is angle between PBS horizontal plane and photon polarization plane as defined in Figure 22. Input photon is monitored by a gating photon, as explained in Chapters 8 and 9, but not shown here.

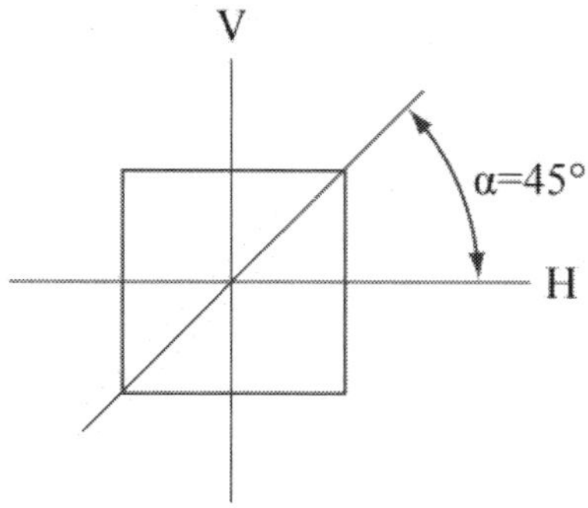

Figure 22.
Front view of PBS.
Linearly polarized photon is directed through PBS with a plane of polarization at an angle of 45 degrees relative to PBS horizontal plane H.

According to QM, photon is a wave packet and its wave packet length is determined by the shape and width of photon source spectrum. It means a wave packet is a mathematical entity with specific wave packet length. But photon cannot have a specific physical length in the direction of travel without getting into conflict with special relativity. Hence, photon cannot interfere with itself, resulting in QM enigma.

In AQM, the photon is not wave packet. It is an elementary particle with the three-dimensional inner structure. The photon physical length is zero. The photon aphysical length is defined by individual photon energy and not by spectrum of its source. The photon aphysical length is non-zero and a multiple of its wavelength. As a result, photon can interfere with itself. Photon interference is a result of aphysical-aphysical interaction of two a-fractions, one of them is the host. Physical energy plays no role. That is how QM enigmas 1 and 2 (see Chapter 3) are explained and resolved. Contrary to QM interpretation, a primary input photon travels along a *specific* trajectory toward PBS.

After passing through PBS, input photon becomes self-entangled by splitting into two a-fractions. One a-fraction, the host, carries the *intact* c-ring with its intact physical energy. Both a-fractions have equally reduced normalized aphysical energy densities, $d_1 = d_2 = 1/2$, where energy density of input photon $d = 1$. Both the host and the a-fraction are connected via c-link.

An a-fraction traveling along Path 1 has the vertical polarization; an a-fraction traveling along Path 2 has the horizontal polarization. Aphysically, the photon travels along both trajectories, Path 1 and Path 2. This explains QM Enigma 4 (see Chapter 3), how a physically indivisible photon can travel along two separate paths.

Combined, both a-fractions constitute a superposition of aphysical linearly polarized structure of input photon. This explains QM Enigma 5, the ontology of the superposition (see Chapter 3).

Here are four fundamental observations:

1. after passing through PBS, the c-ring physical substance adopts aphysical linear polarization of the host;
2. distribution of the aphysical density, d_1 and d_2, between Path 1 and Path 2 is *deterministic* and defined by the ratio $d_1 : d_2 = \sin^2 a : \cos^2 a$;
3. ratio of normalized aphysical density, d_1 and d_2 defines probability of photon c-ring, with its physical substance, to travel along either Path 1 or Path 2;
4. AQM expands areas of determinism and causality.

Both host and a-fraction travel at the speed of light in free space. It can be shown more compellingly in Figure 23, where both host and a-fraction travel in opposite directions, separating from each other with the double speed of light relative to their frames of self-reference. Such statement is not in conflict with special relativity because aphysical energy is not part of special relativity.

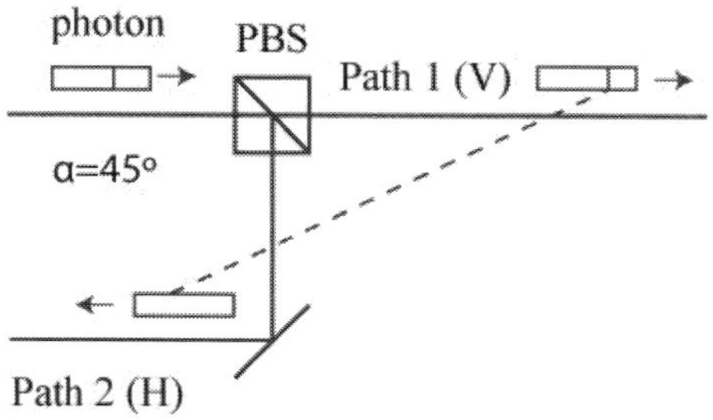

Figure 23.
In this experimental layout, host and a-fraction travel in opposite directions, each with the speed of light.

Let us install two detectors, D_1 and D_2 in Path 1 and Path 2, as shown in Fig. 24.

With a single photon sent through PBS, one should expect that only one of two detectors will "Click" confirming that photon is registered. Outcomes with "Two Clicks" or "No Clicks" are excluded. When D_1 or D_2 clicks, the path taken by the photon physical substance (the host) is identified. The decision is made *at PBS* and *not at detection.* Here we deal with two events: a single photon passing through the PBS (event 1) and the "Click" at one of detectors (event 2). To say that the decision is made at the detection means that the present (event 2) influences the past (event 1).

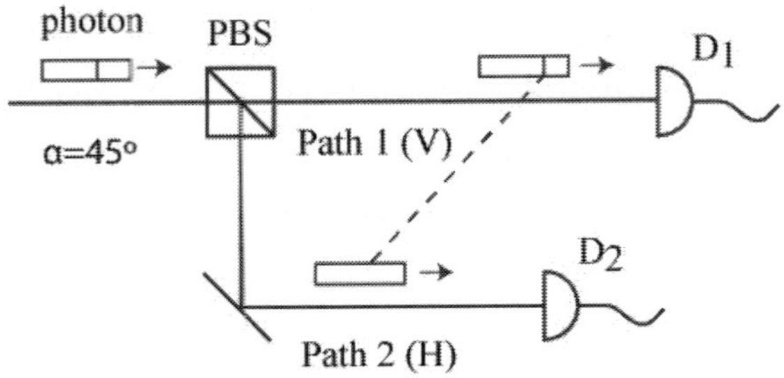

Figure 24.
Two detectors D_1 and D_2 are installed in Path 1 and Path 2, respectively.

Let us remove one detector (D_1) and leave in place only detector D_2 (see Figures 25 and 26).

With a single incoming photon and a single detector, we have one of two possible outcomes: either "Click" or "No Click" at detector D_2.

In case of "Click", we know that the host is traveling along Path 2. An a-fraction is instantaneously recalled by the host from Path 1. The photon acquires the full-fledged status before it is destroyed and absorbed by detector D_2 (see Figure 25).

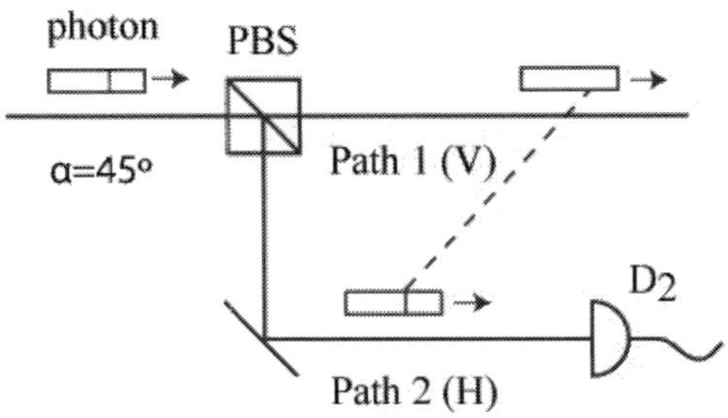

Figure 25.
The "Click" outcome at detector D_2

n the case of "No Click", we now know that the host is traveling along Path 1. A-fraction, after interacting *aphysically-physically* with detector D_2, is instantaneously recalled by the host. The photon is reconstructed to its full-fledged status and continues its travel along Path 1 (see Figure 26).

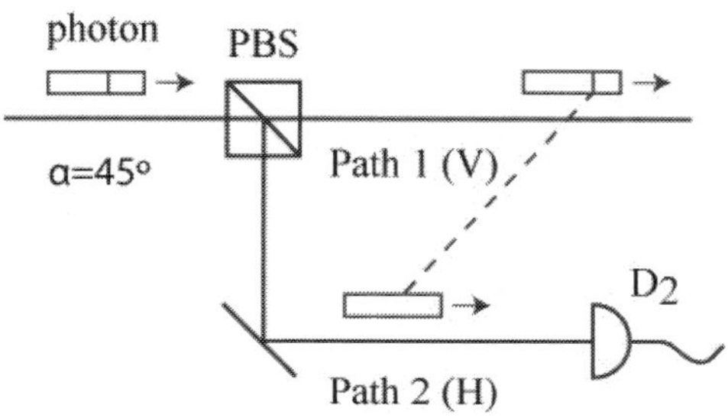

Figure 26.
The "No Click" outcome

As I observed in my paper [2], this is the case of photon *non-physical* interaction.

Again, the "No Click" case is real. It provides as much experimental evidence as the "Click" case. It confirms the conclusion on where the decision is made. The decision is made *at PBS*. The "No Click" causes "the collapse of the superposition" in QM terminology. Instantaneous action, regardless of separating distance, and instantaneous collapse do not contradict special relativity. This is the explanation of QM Enigmas 6, 7, 8 and 9 (see Chapter 3).

As one can see, AQM explains QM Enigmas in the simplest quantum optics experiment, such as PBS with a single-photon source. And, no mathematical formalism is needed.

Chapter 6

Conceptual Three-Level Beam Splitter Experiment as an Example of Aphysical Quantum Optics

A conceptual layout for a three-level polarizing beam splitter with linearly polarized photons is shown in Figure 27.

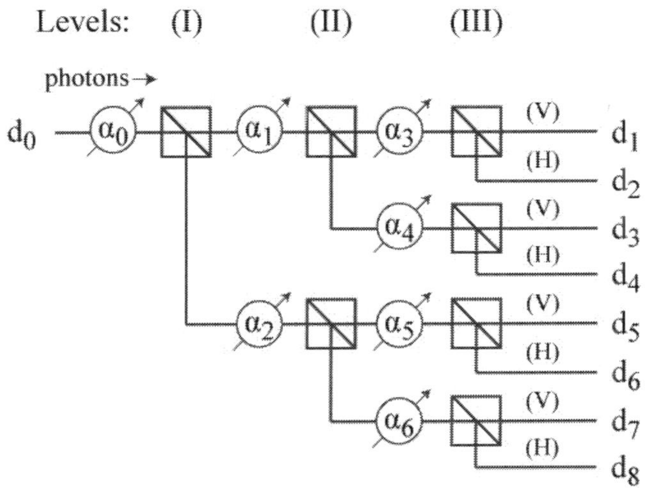

Figure 27.
Experimental three-level (I,II,III) beam splitter layout with linearly polarized signal photons

In comparison with a single beam splitter, the three-level layout is more complex, demonstrating in depth incredible versatility and beauty of photon properties discovered in AQM.

We know the orientation of planes of polarizations, (H) and (V), for both a-fractions of self-entangled photon exiting the beam splitter (Level I), therefore we can preset entry angles for the next stage of beam-splitters (Level II) (see Figures 27 and 28).

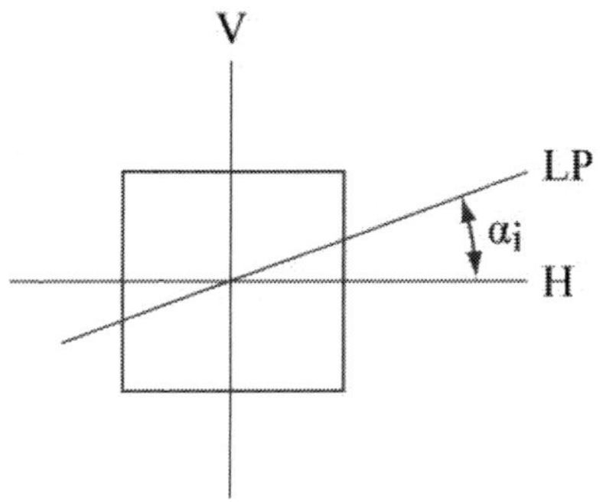

Figure 28.
Front view of a beam splitter.
H and V are horizontal and vertical planes.
LP is a photon linear polarization plane

Of course, in place of mechanical positioning of beam splitters, one can use electro-optics modulators but the purpose of this study is conceptual rather than practical experimentation.

According to QM, a photon, after passing this maze of beam splitters, is in the superposition of eight quantum states. The quantum state vector is defined as,

$$\Psi = \sum_i \alpha_i \Psi_i$$

where $\sum_i |\alpha_i|^2 = 1$ and $i = 1,2..8$.

On the other hand, according to AQM, the three-level system with polarizing beam splitters is a quantum optics system for the distribution of aphysical energy density in a predictable deterministic way, from a single input down to eight separate outputs, *with no role* for the physical substance to play. The physical substance is being carried along by its host. This sounds incredible.

Each individual full-fledged input photon is converted at the exit into a self-entangled photon with eight a-fractions; only one of them, the host, carries the physical substance. *The selection of a-fraction as a host is an intrinsically irreducible random probabilistic process.* The probability for a given a-fraction chosen to be a host and carrying the physical substance is equal to its normalized aphysical energy density.

Here is a simple but accurate calculation of normalized aphysical energy density distribution among eight output a-fractions, with input photon aphysical density assigned as $d_0 = 1$ (as shown in Figure 29). To simplify, let us introduce the following notation, $\sin^2 \alpha_n \equiv s_n$, $\cos^2 \alpha_n \equiv c_n$, where $n = 1,2...8$. This gives us:

$$d_1 = c_0 c_1 c_3$$

$$d_2 = c_0 c_1 s_3$$

$$d_3 = c_0 s_1 c_4$$

$$d_4 = c_0 s_1 s_4$$

$$d_5 = s_0 c_2 c_5$$

$$d_6 = s_0 c_2 s_5$$

$$d_7 = s_0 s_2 c_6$$

$$d_8 = s_0 s_2 s_6$$

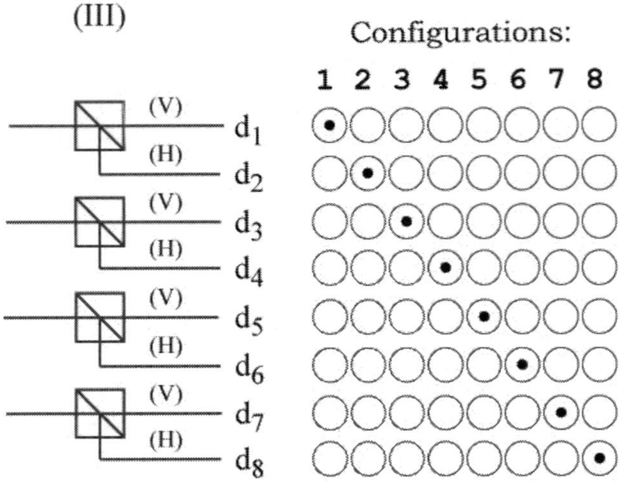

Figure 29.
Eight output quantum state configurations of photon self-entanglement in the three-level PBS experiment.
Only level III is shown.

⊙ the host

○ a-fraction

Figure 29 shows the output of Level III beam splitters (shown in Figure 27) and the eight possible configurations of photon self-entanglement. Let us assume that our system is in free space. Let us further assume that we set up eight reflecting mirrors in eight arbitrary directions from each of eight outputs of the system, as shown in Figure 30.

As a specific example, configuration 5 of a self-entangled photon is selected from Figure 29 and presented in free space in Figure 31.

Three output parameters are predictable and deterministic for each individual a-fraction: aphysical energy density, aphysical fractional spin, and plane of polarization. As mentioned above, the selection of a-fraction as the host is an intrinsic probabilistic process.

Free space is basically transparent to photon. Seven a-fractions and the host can travel intact for millions or billions of years.

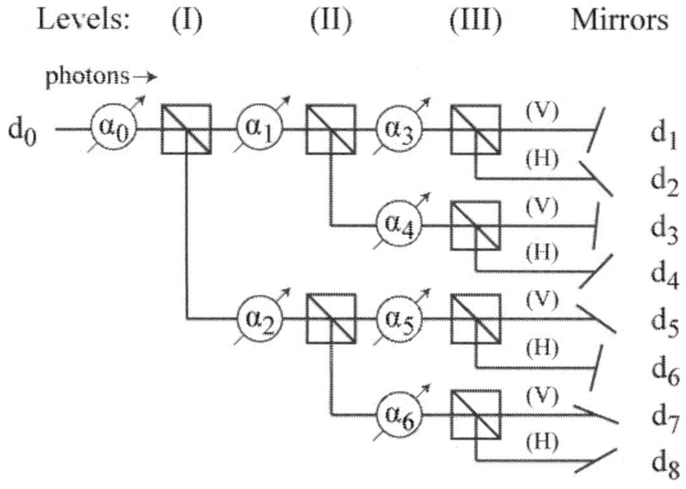

Figure 30.
Reflecting mirrors installed in eight arbitrary directions at the eight outputs of the three-level system. (V) – vertical polarization, (H) – horizontal polarization.

Elementary consciousness, which resides in the c-ring of the host, is aware of the status of each of the seven a-fractions via connecting c-links. If the host encounters a physical absorbing object (not an optical one), elementary consciousness proceeds with an instantaneous reconstruction of the photon into the full-fledged photon, regardless of the separating distance, before proceeding with a physical-physical interaction with the object.

In case an individual a-fraction encounters a physical object, it is instantaneously recalled by the elementary consciousness of the host regardless of the separating distance, resulting in partial reconstruction.

Let us now install eight output detectors, as shown in Figure 32. Probability of detector D_i to click is equal to normalized individual output density d_i ($i=1,2...8$).

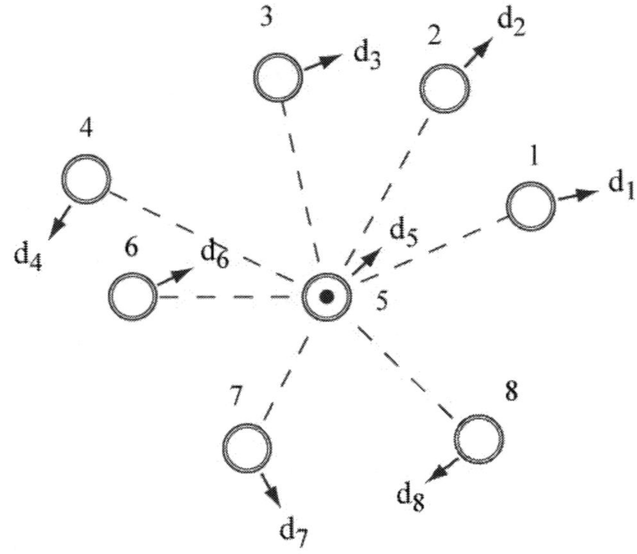

Figure 31.
An example of a self-entangled photon in the configuration #5 in free space.
d_n - aphysical energy density ($n=1,2...8$) of individual a-fractions and host; arrows are directions of travel.

According to AQM, "No Click" is a result of aphysical-physical interaction of an a-fraction with a physical detector. Science has not yet developed a technique for detection of aphysical-physical energy interactions. This is still in the future.

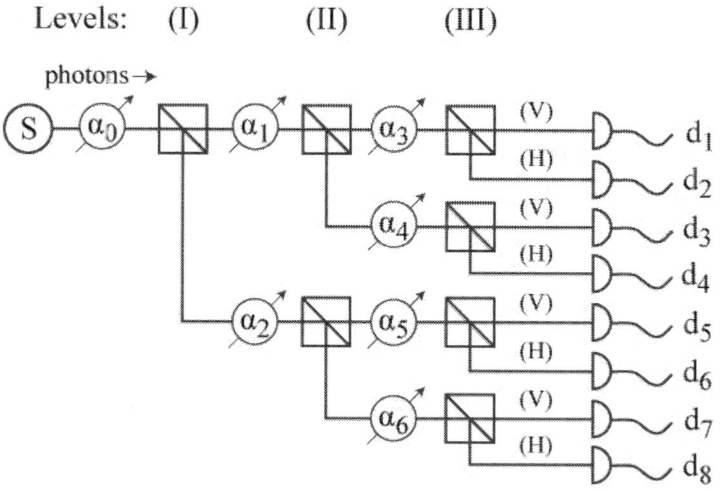

Figure 32.
Three-level system with output detectors installed.

For non-believers, I can offer simple proof. Please connect any two outputs to a Mach-Zehnder interferometer with a variable phase shifter, as explained in Chapter 8. Of course, the planes of polarization must be adjusted accordingly. The easiest way would be to use any pair of outputs, one with vertical and another with horizontal polarization. Then the polarizations would be in the correct relation for a Max-Zehnder interference experiment. But in principle, one can produce an interference pattern with any two outputs.

Comment: Figure 33 shows a higher order photon self-entanglement.

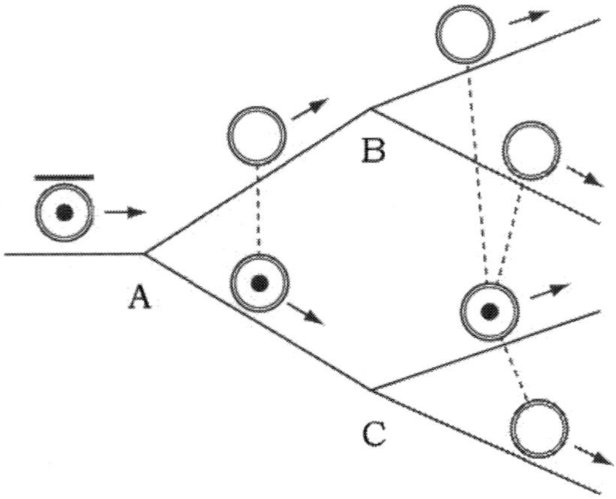

Figure 33.
An example of two level system with a higher order photon self-entanglement shown at point B, where a-fraction is split in two a-fractions and self-entangled. Both are c-linked to the host. General notation:

 the full fledged photon

 the host

 a-fraction

Chapter 7

The Heisenberg Uncertainty Principle. Photon Coherency Length in QM interpretation

As a probabilistic science, QM formalism deals with assemblies of photons and not with individual photons.

In quantum optics, it is customary to describe photon as a wave packet with all of its physical parameters assigned, such as angular velocity ω, wavelength λ, energy $E=\hbar\omega$, momentum $p=E/c$, and spin *S*.

The photon wave packet length is equated to the coherence length.

As established in quantum optics (or, generally, in optics), the coherence length of photon source is determined by its spectrum:

$$L_c(\omega)=k\frac{\lambda_0^2}{\Delta\lambda}, \qquad (7\text{-}1)$$

where $\Delta\lambda$ is the full wavelength bandwidth, λ_0 is the central wavelength of photon source spectrum, and *k* is *k*- factor in the range of 0.66 to 1.33 (see Figure 34).

The relation (7-1) is approximate. Ambiguity is due to various definitions of the full bandwidth and by the details of the source spectrum.

As one can see later, for my purpose, such ambiguity of the k-factor is irrelevant.

In quantum optics [16], it is assumed that k-factor is equal to 1 and the coherence length is presented as

$$L_c(\omega) = \frac{\lambda_0^2}{\Delta\lambda}, \qquad (7\text{-}2)$$

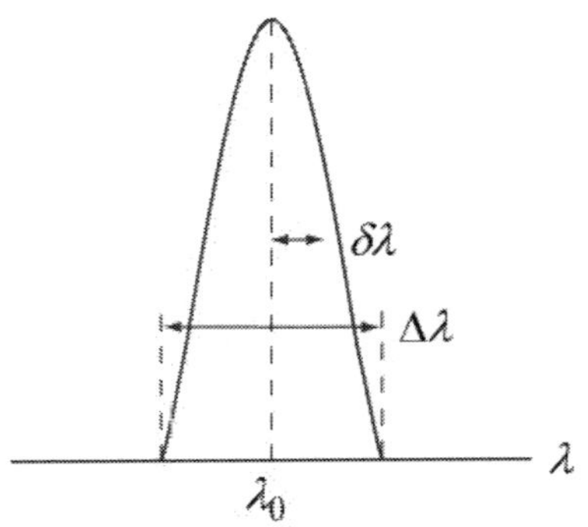

Figure 34.
Photon source spectrum:
$\delta\lambda$ – half bandwidth at 50% level. $\Delta\lambda$ – full bandwidth at 0% level. $\Delta\lambda \approx 3\delta\lambda$

If one knows the source spectrum, one can calculate approximately its coherence length $L_c(\omega)$ (notation "ω" stands for source spectrum).

Another way to determine the coherence length experimentally is by measuring the length of interference $L_c(I)$ using the Mach-Zehnder interferometer (notation "I" stands for interference). Figure 35 shows a simplified Mach-Zehnder layout.

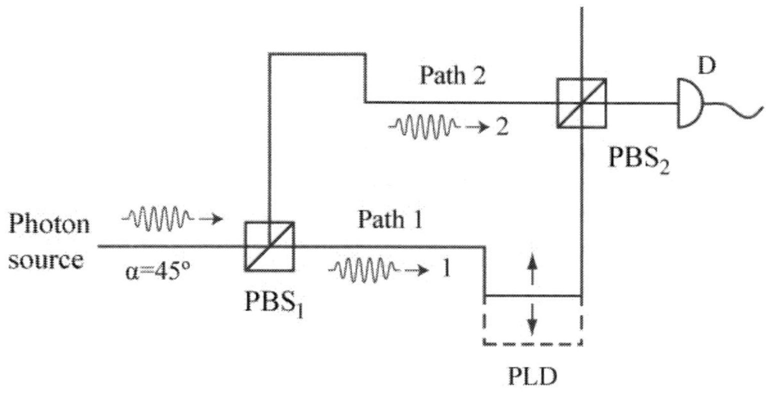

Figure 35.
Mach-Zehnder interferometer experiment in QM interpretation.
PBS_1 and PBS_2 – polarizing beam splitters;
PLD – adjustable and calibrated pathlength difference device;
D – detector. (H) and (V) are horizontal and vertical polarizations

The coherence length $L_c(I)$ is measured as the path length difference between points A and C, where visibility V is zero (see Figure 36). Visibility is defined as

$$V = \frac{N_{\max} - N_{\min}}{N_{\max} + N_{\min}} \tag{7-3}$$

where N is the fringe count at detector D.

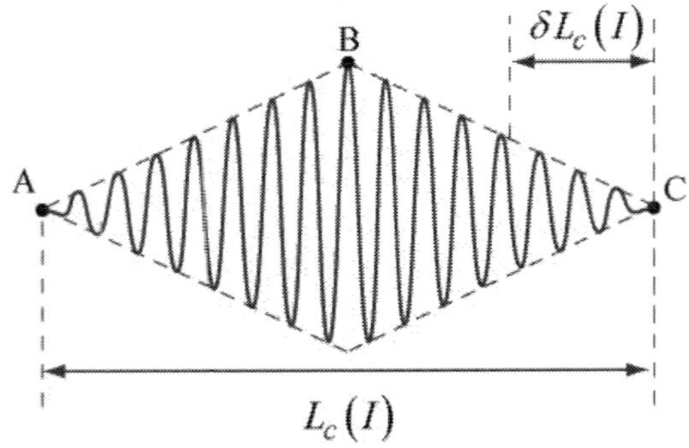

Figure 36.
An example of measurement of photon coherence length $L_c(I)$ with Mach-Zehnder interferometer

As shown in Figure 36, by adjusting path length difference (PLD) between Path 1 and Path 2, the "discrepancy" phase φ is gradually accrued between wave-packets on Path 1 and Path 2. At point A $\varphi = -\pi$; at point B $\varphi = 0$; and at point C $\varphi = \pi$ (and all this is referencing figure 36 – shouldn't it say so?). The phase difference between points A and C is equal 2π, or one wavelength.

In quantum optics, experimental value of the coherence length $L_c(I)$ is equal to path length difference between points A and C, corresponding to discrepancy phase 2π, or one wavelength.

The shape of the interference pattern, as shown in Figure 36, is "rhombus-like" (also see experimental evidence in [14, 15]). For monoenergetic photons, the shape of the interference pattern is indistinguishable from the rhombus. QM has no ontological explanation why the interference pattern has this shape. In contrast, AQM provides a straightforward ontological explanation.

In quantum optics, there are two different definitions of the coherence length. One, $L_c(I)$, as described above, is based on experimental measurements of the interference pattern length; and the second one, $L_c(\omega)$, is based on the spectrum of the photon source and the Heisenberg uncertainty principle.

What is the relation between $L_c(\omega)$ and $L_c(I)$? One derives this directly from the Heisenberg uncertainty principle as follows:

$$\delta t \delta E \geq \hbar/2 \qquad (7\text{-}4)$$

where

$$\delta t = \delta l/c,$$

$$\delta E = 2\pi\hbar\delta v,$$

$$\delta v = \left|\delta\lambda/\lambda^2\right| c,$$

$$\Delta\lambda = 3\delta\lambda$$

Putting it all together, one obtains

$$L_c(\omega) = \lambda^2/\Delta\lambda .$$

As a result, we can conclude that according to QM, both definitions $L_c(\omega)$ and $L_c(I)$ are approximately equal:

$$L_c(\omega) \approx L_c(I) \qquad (7\text{-}5)$$

The relation is linear as shown in Figure 37. It is also assumed that the size of Mach-Zehnder interferometer is larger than the coherence length.

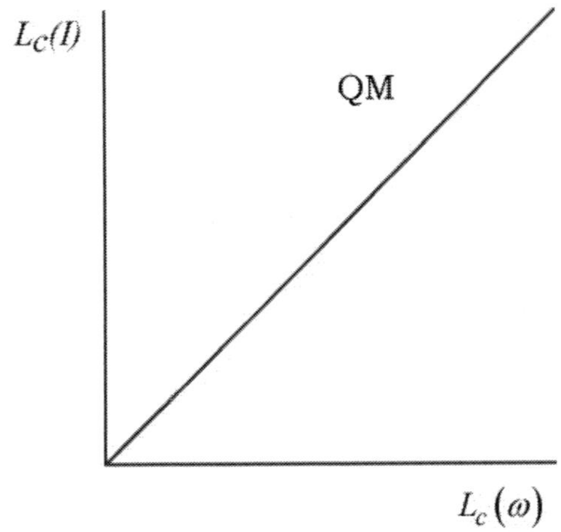

Figure 37.

Linear relation between photon coherence length $L_c(\omega)$ and photon interference pattern length $L_c(I)$, according to QM interpretation.

The relation (7-5) is so obvious to quantum optics physicists that they do not bother to verify it experimentally. Who would think to verify the Heisenberg uncertainty principle? As the saying goes, “so far, nobody has proved quantum mechanics wrong.”

AQM proves that the relation (7-5) is invalid.

Chapter 8

AQM Theoretical Prediction and Proposed Conceptual Experiment for Measurement of Photon Aphysical Length

The photon, as an elementary particle, consists of three components: the physical energy c-ring, the aphysical energy cylinder, and the elementary consciousness residing in the c-ring.

In the case of the linearly polarized photon, physical energy bundle is split into two equal sub-bundles, each spinning in opposite direction with equal Compton angular velocity. The aphysical energy cylinder is also split equally in two aphysical sub-cylinders, each spinning in opposite directions. To be consistent with special relativity, the length of the photon physical energy bundle, the c-ring, is zero in the direction of photon travel. In contrast, aphysical energy is not subject to special relativity, and photon aphysical cylinder length L_a is not affected by relativistic effects and remains constant. *The normalized length of the aphysical cylinder* L_a/λ *is the AQM photon aphysical fundamental constant.*

Can we determine a value of the photon aphysical constant experimentally? Yes, by performing a simple quantum optics experiment using Mach-Zehnder interferometer. Then why has this not been done already? The reason the photon aphysical constant has not yet been discovered is due to QM dogmatic mindset, which is an overwhelmingly paralyzing factor in quantum physics. The experiment itself is very simple. It can be done in college undergraduate labs.

A simplified concept of the Mach-Zehnder experiment layout is shown in Figure 38. However, in this case the experiment is based on *AQM interpretation and AQM guidance.*

The actual layout includes a single-photon source using the down conversion for production of pairs of correlated linearly polarized photons, with one photon as gating photon and another as signal photon. As shown in paper [16], the layout may include a set of optical filters to vary the coherence length $L_c(\omega)$.

The more elaborate experimental layout is described in detail in Chapter 9, with full credit given to quantum optical physicists who published paper [16], although some technical details are different, as discussed later. My conceptual experimental layout uses a strictly linear variable path difference device (PLD), which in this case is important for accurate measurements of the AQM photon fundamental aphysical constant.

The experiment [16] is designed for training undergraduate students in quantum optics experimentation. *It has nothing to do with the purpose of my experiment.*

The purpose of my conceptual experiment is as follows: (1) to demonstrate scientific validity of the AQM photon model, (2) to show how to measure the photon aphysical fundamental constant L_a/λ, and (3) to demonstrate a breakdown of the Heisenberg uncertainty principle as applied to individual elementary particle interactions.

In quantum optics, the concept of the wave packet is scientifically invalid. The AQM photon is no longer described as wave packet.

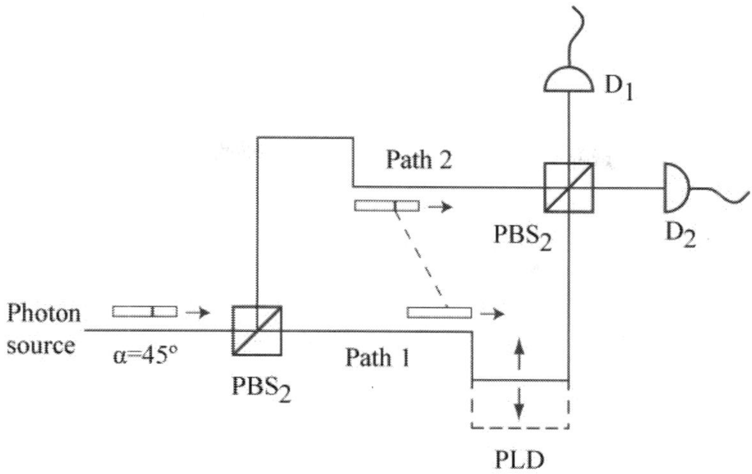

Figure 38.
Simplified experimental layout for measurement of photon aphysical length.

PLD – adjustable and calibrated path length difference device. A linearly polarized photon (signal photon) is directed to PBS_2 at 45 degrees

As shown in Figure 4, the photon has three-dimensional cylindrical inner structure consisting of the physical energy c-ring with zero length, embedded into the hollow aphysical energy cylinder (AC) of non-zero length. Both the c-ring and the AC have walls with zero thickness, the same Compton radius r and the same Compton angular velocity ω. The circumferential linear velocity of the photon structure is equal to the speed of light; wavelength λ is defined as $\lambda = 2\pi\hbar c / E_p$, where $E_p = \hbar\omega$ is photon physical self-energy, and $E_a = E_p / U$ is photon aphysical self-energy; wavelength λ is equal to the circumference of the photon cylindrical inner structure.

A linearly polarized photon (a signal photon) is sent into PBS_1 at an angle of 45 degrees. After passing through PBS_1, the photon becomes self-entangled. It is split into two a-fractions with equal aphysical energy density: one a-fraction with vertical polarization travels along Path 1 and another a-fraction with horizontal polarization travels along Path 2. With 50/50 probability, only one a-fraction becomes the host and carries the photon physical substance (the c-ring). Both the host and the a-fraction are connected by the c-link. The polarization of the c-ring always conforms to the host polarization. It is important to emphasize that *the physical substance (the c-ring) always acquires (follows) the host polarization.*

Location of the physical substance along the host aphysical cylinder is called the position parameter (PP). Figure 39 shows two possible situations: (a) host in Path 2; and (b) host in Path 1.

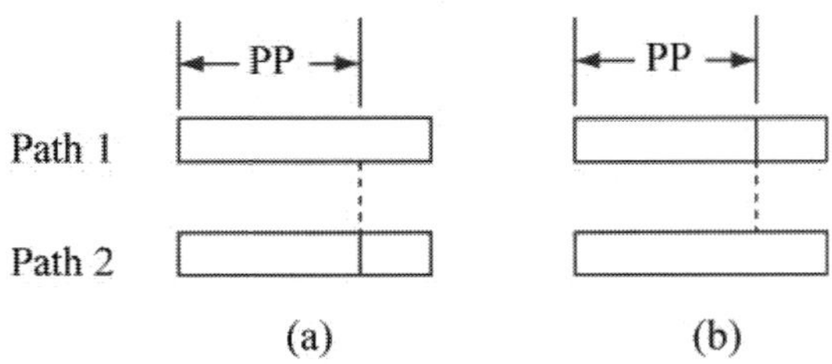

Figure 39.
Two possible situations:
(a) host in Path 2; and (b) host in Path 1.

The position parameter is a source of irreducible quantum randomness. In principle, each photon is unique. One can never find two photons with the same value of the position parameter.

♦ ♦ ♦

Postulate 43. *The Position Parameter Distribution in Large Assembly of Monoenergetic Photons.*

The position parameter is quantum irreducible random parameter. It is unique to each individual photon. In a large assembly of monoenergetic photons, the position parameter is distributed uniformly along the length of the photon aphysical cylinder.

♦ ♦ ♦

Conceptually, one can divide an assembly of entangled photons into two sub-assemblies: one sub-assembly with the host on Path 1 and second sub-assembly with the host on Path 2. Both assemblies are symmetrical and probabilistically equal, with uniform density distribution of the c-rings along the aphysical cylinder (see Figure 40). For purposes of discussion, it is sufficient to consider only one sub-assembly, in this case sub-assembly (a) in Figure 40.

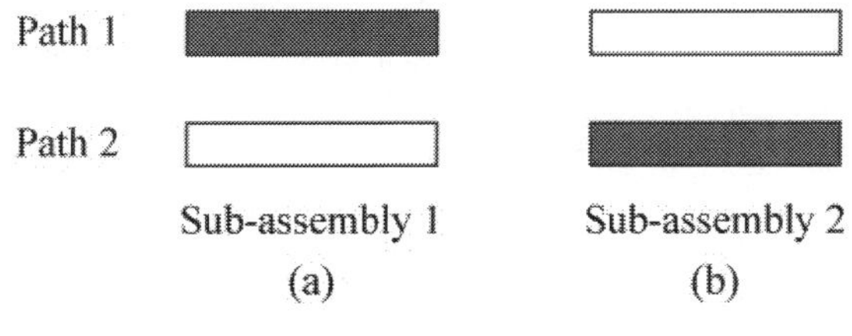

Figure 40.
Two sub-assemblies of monoenergetic signal photons.

Expected interference pattern at detector D_1 or D_2 is shown in the Figure 41. Now it is clear why the interference pattern with monoenergetic photons has a strictly rhombus-like shape. *This is the AQM ontological explanation.*

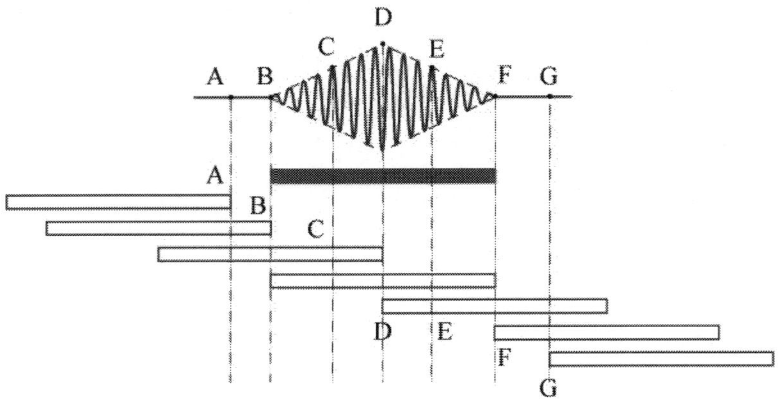

Figure 41.
The interference pattern at detector D_1 or detector D_2

The interference pattern is the result of aphysical-aphysical interaction of the host and a-fractions at PBS_2, where the c-ring physical substance has no role to play. In such interaction the physical substance is taken "for a ride." The physical substance is statistically superimposed on aphysical-aphysical interference pattern, allowing us detect and measure it.

Signal and gating photons of high coherence, combined with adjustable and calibrated path length difference (PLD), allow us measure and determine *the photon aphysical length.* The process of the interference pattern build-up is shown in Figure 41, where one varies the path length difference of one a-fraction relative to another. In mathematical terms, the process is known as *convolution.*

As mentioned previously, in an assembly of photons, individual c-rings are randomly positioned along the aphysical cylinder length but statistically uniformly distributed.

To verify this, let us now assume that the c-ring is always located in the same position within the aphysical cylinder for all photons (for example, at the center of aphysical cylinder). The expected interference pattern would then be rectangular, as shown in Figure 42, conflicting with experimental observation.

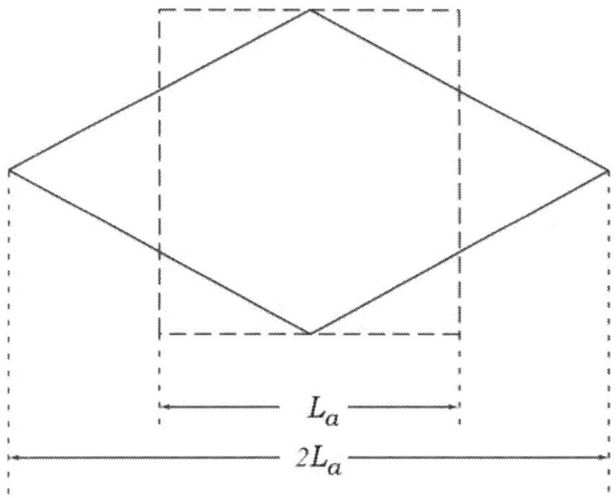

Figure 42.
Shape of the interference patterns:

(a) "rhombus" based on the assumption that the distribution of the position parameter (the location of c-ring) is uniform for a large assembly of monoenergetic photons, or

(b) "rectangular" based on the assumption that the position parameter (the location of c-ring) is in a fixed position in the aphysical cylinder.

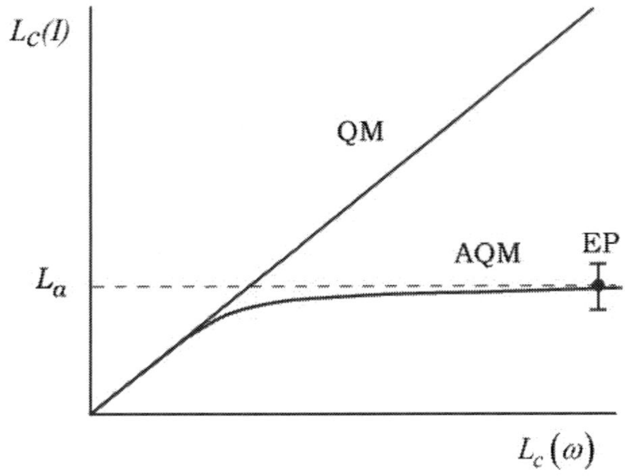

Figure 43.
Experimental expectation of Quantum Mechanics versus Aphysical Quantum Mechanics

In conclusion, as shown in Figure 43, QM and AQM definitions of the coherent length are no longer identical. They diverge. As shown in Figure 43, AQM interpretation allows one experimentally determine the photon aphysical length.

I was able to derive only one single experimental point EP from available and unrelated experimental data (see Chapter 8).

λ_c	- photon wavelength equal to the circumference of the aphysical cylinder;
$L_c(I)$	- photon interference pattern length;
$L_c(\omega)$	- coherence length of photon source, same as the length of photon wave packet;
L_a	- photon aphysical length;
L_a / λ_c	- photon fundamental aphysical constant.

For example, for λ_c equal to 0.6 μm, experimentally estimated photon aphysical length is equal to 90 μm (+100 μm, - 30 μm)

.

Chapter 9

In Search of Experimental Evidence for the Confirmation of AQM Photon Model

There is a clear-cut conflict between the Heisenberg uncertainty principle and AQM definition of the photon aphysical length.

While searching through numerous published quantum optics papers, I found, to my incredible luck, a paper published in American Journal of Physics in 2005 by seven authors: E.J. Galvez, C. H. Holbrow, M. J. Pysher, J. W. Martin, N. Courtemanche, L. Heilig, and J. Spencer, with the title "Interference with Correlated Photons: Five Quantum Mechanics Experiments for Undergraduates". Please notice the term, "for undergraduates." [16]

Among the experimental data, there is one unique experimental point revealing that QM is wrong. Therefore, the often-quoted statement, "QM has never been proven experimentally wrong", is a myth. In fact, the mere existence of so many enigmas and numerous interpretations is sufficient evidence that QM is a fundamentally invalid quantum theory. The experimental data also shows that the Heisenberg uncertainty principle has limits of scientific validity. It works only statistically and is not applicable to individual quantum entities and individual quantum processes.

The experimental data is the first confirmation of scientific validity of the AQM photon model, and this data is sufficient to estimate the photon aphysical length. *The first fundamental aphysical constant is experimentally confirmed.*

An experimental layout, as presented in the paper [16], is reproduced in Figure 44, where the photon source is 457.9 nm argon-ion laser. This type of laser is known for its high coherence in the range of 1 to 10 meters, more than sufficient enough for our purpose. In fact, the coherence of even 10 cm would be sufficient.

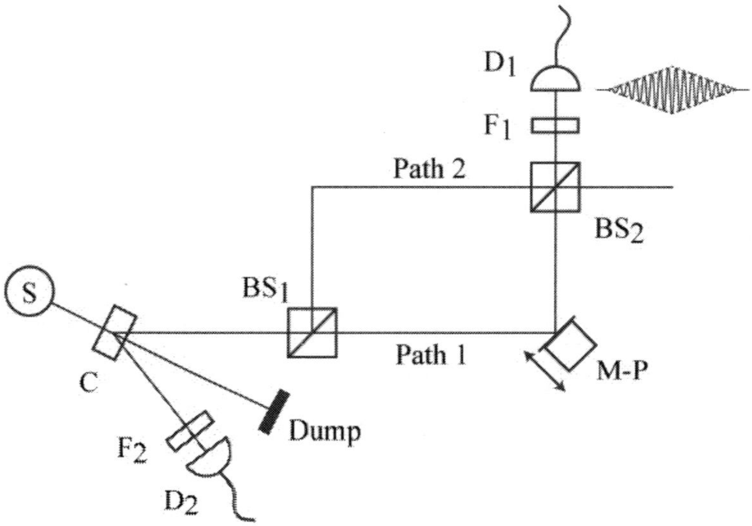

Figure 44.
Experimental layout from Ref. [16]

A beta-barium-borate down-conversion crystal (C) is used to produce pairs of down-conversion twin photons which leave the crystal at ± 3 degrees relative to the pump-beam axis.

Both down-conversion photons have equal (or very close to equal) wavelength, λ=915.8 nm. PBS_1 and PBS_2 are polarizing beam splitters. M-P is a micrometer-piezo-driven mirror used for adjusting the path length difference (PLD). Two avalanche photodiodes, D_1 and D_2, are installed to register signal photons and gating photons, respectively.

The important parameter, called "visibility", is a measure of completeness of interference. It is defined as,

$$V = \frac{N_{max} - N_{min}}{N_{max} + N_{min}}$$

where N_{max} and N_{min} are the maximum and minimum counts in the interference pattern by detector D_1.

At PLD equal to 0 (both optical paths are equal), V=1 (full interference).

PLD equal to $l/2$ is the cut-off point where interference disappears, and l is assumed to be the length of the wave packet, as per QM interpretation.

Two pathband filters, F_1 and F_2, are installed in front of detectors D_1 and D_2, respectively. Filter F_1 has pathband equal to 10 nm. Filter F_2 is a set of interchangeable filters, to be installed one-at-a-time from one experiment to the next one, with pathbands of 10 nm, 1 nm and 0.1 nm, respectively.

A unique feature of the experimental setup is the filter with very narrow pathband of 0.1 nm. Without such filter, the experiment would not be conclusive to verify AQM photon model and experimentally determine the photon aphysical length.

Interchangeable filters in front of detectors D_1 and D_2 determine the photon coherence length as follows:

F_1	F_2	Coherence length
10 nm	10 nm	168 μm
10 nm	1 nm	924 μm
10 nm	0.1 nm	8400 μm

Here, as the authors of the paper [16] write:

> "As a dramatic demonstration of the effect of the magnitude of the coherence length, we placed a 0.1 nm filter in front of the idler detector with l = 144 μm. The new filter increased the coherence length to 8400 μm, and, and as can be seen by the squares in Figure 3 (b) the fringes reappeared with a visibility of V = 0.59 ± 0.04."

The Heisenberg uncertainty principle is no longer applicable in this case, and QM is proven wrong experimentally by only one but conclusive experimental point, as follows:

$$V = 1 \qquad \text{at } l = 0$$

$$V = 0.59 \pm 0.04 \qquad \text{at } l = 144\ \mu\text{m}$$

By linear extrapolation, one can calculate the approximate interference pattern length of 320 μm. According to QM, the interference pattern length must be equal to the source coherence length. In the experiment, the source coherence length is 8400 μm as compared to the interference pattern length of 320 μm. It shows a *huge* discrepancy between QM and AQM (with factor 56). AQM prediction is proven correct—*the photon aphysical length L_a is established in approximation.* After normalization by photon wave length, $\lambda = 915.8$ nm, a normalized length of the aphysical cylinder is approximately:

$$L_a/\lambda = 360$$

However, there is one important observation. As one can see in Figure 45, the authors use a peculiar method, common in quantum optics, for adjustment of path length difference by driving mirror M-P. If we remove PBS_2, then the mirror movement produces no change in path length difference, as shown in Figure 46. Such method results in a non-linear adjustment of the path length difference. The nonlinearity tends to *exaggerate* the value of the photon aphysical length. For example, the actual path length difference might be closer to 72 μm (and not 144 μm), then the photon normalized length of aphysical cylinder is reduced by factor 2, becoming:

$$L_a/\lambda = 180$$

Such correction makes AQM case even stronger.

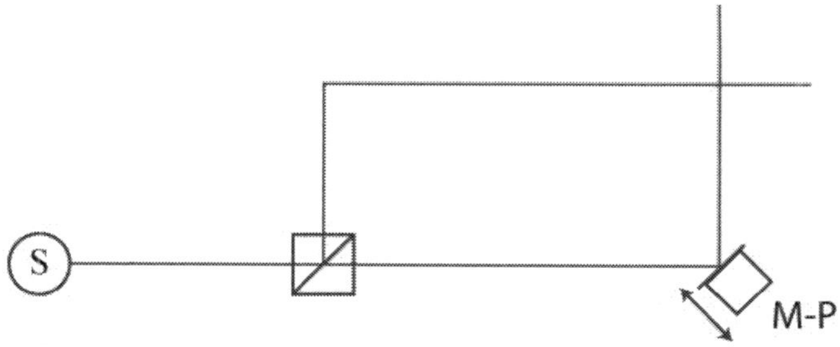

Figure 45.
Moving mirror M-P for adjustment of
path length difference

The way to adjust the path length is shown in Figure 46, where one achieves a strictly linear relation between the mechanical adjustment of the PLD and the path length difference over any required range. Some quantum optics physicists use this method.

In the absence of more accurate experimental data, in my estimation, the normalized photon aphysical length, L_a/λ, is in the range of 100 to 400, with 150 being the most probable value.

This is 56 times smaller than the value of the photon coherence length predicted by QM.

The authors [16] are oblivious to such enormous discrepancy! They stumbled into a "mother lode" without realizing it. But they are given full credit here.

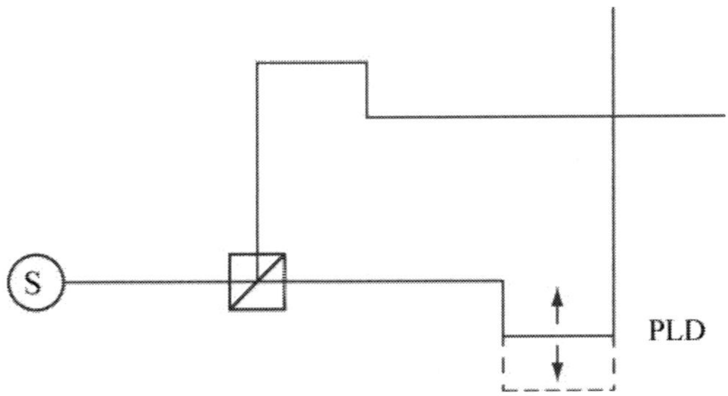

Figure 46.
PLD – linear adjustment of the path length difference

Let us take an example of a photon with energy in the middle of visual spectrum with wavelength $\lambda = 0.6$ μm. Thus, photon diameter is 0.18 μm, and the aphysical length is 90 μm. Yes, 90 μm! It is no longer a surprise why the photon aphysical energy cylinder plays a predominant role in the aphysical domain. It also confirms the importance of PLD linearity.

Finally, QM has been proven experimentally wrong. The AQM photon model is experimentally confirmed, although with an approximate value of a single experimental point. The first fundamental aphysical constant is experimentally confirmed.

Henceforth, the normalized photon aphysical length constant joins the class of other fundamental constants, such as the speed of light in vacuum c, Planck constant h, elementary electric charge e, and others.

Comment: The experimental layout presented in Figure 44 shows how one can get around the restrictions imposed by the Heisenberg uncertainty principle. As the pathband of filter F_2 is further reduced, energy of the gating photon is defined more accurately, and thus its arrival to detector D_2 becomes more uncertain. But this does not influence the timing definition for the signal photon because once the gating photon is registered by detector D_2, the exact timing of arrival of the signal photon to detector D_1 is known.

Chapter 10

Conclusion

"All these fifty years of conscious brooding have brought me no nearer to the answer to the question 'What is light quanta?' Nowadays every physicist thinks he knows it, but he is mistaken." (Albert Einstein, Letter to M. Besso, 1951).

♦ ♦ ♦

Postulate 44. *The complete definition and properties of the photon based on Aphysical Quantum Mechanics.*

The photon is the force of electromagnetism and the mediator of electromagnetic interaction.

The photon is the quantum of inverted electromagnetic energy (in fact, electromagnetostatic energy).

The photon can be understood to the full extent only on the basis of the AQM quantum reality. The AQM quantum reality includes three fundamental categories: physical, aphysical, and elementary consciousness. The photon has been hardly explored in quantum mechanics, quantum optics, and particle physics. Fundamental definition and properties of the photon are explained and dramatically expanded in AQM .

The inner structure of photon consists of three constituents: the physical energy c-ring, the aphysical energy cylinder, and elementary consciousness. The normalized aphysical energy *cylinder length L_a/λ is a fundamental photon aphysical constant. Its value is found experimentally in some approximation so far.*

The photon is a boson traveling in free space with the speed of light. It is stable and can travel in the universe for billions of years, gradually reducing its energy due to space expansion. The photon has spin S = +1 for the right-handed circular polarization and S = -1 for the left-handed circular polarization (in $\hbar$ units). In optical energy range, photon spin can have any value in the range of [-1, +1], as experimentally demonstrated in quantum optics.

In contrast to the fermion, where only one half of physical energy contributes to its spin, producing spin -1/2, in the photon (or generally, the boson) the total physical energy contributes to spin producing spin -1.

In the optical energy range, the c-ring physical energy bundle can be split into two sub-bundles, each spinning with equal but opposite Compton angular velocity. That explains why the optical photon has spin values in the range of [-1, +1]. If the physical energy bundle is evenly split, spin is also split in two equal sub-spins that oppose each other, resulting in zero spin (hidden spin). In such case, photon acquires the linear polarization.

The aphysical energy cylinder imitates dynamics of the c-ring. When the c-ring splits in two sub-bundles of physical energy, the aphysical energy cylinder also splits in two aphysical sub-cylinders.

The origin of photon physical self-mass $h\omega / c^2$ is the spinning c-ring. The nature of photon physical self-mass is inverted electromagnetostatic energy.

The photon has neither intrinsic momentum nor intrinsic kinetic energy in any frames of references, including its own frame of self-reference (v = c). It is a counterintuitive statement.

The photon "inverted" electromagnetostatic energy is contained in the c-ring as physical energy bundle and in the aphysical cylinder as aphysical energy with uniform distribution of both energy densities along the circumference. In compliance with special relativity law, the c-ring has zero cross-section.

The photon's cylindrical structure spins around its axis at Compton angular velocity $\omega = c / r$, where r is Compton radius. Photon wavelength $\lambda = 2\pi r$ is the circumference of the photon cylindrical inner structure.. Photon physical energy is $E_P = \hbar\omega$ and aphysical energy $E_a = E_P / U$. Zero cross-section of c-ring means that the cross-section of photon-photon interaction is zero. This explains why the universe is transparent to traveling photons.

The photon is a particle, never a wave. What is called photon wavelength is in fact the circumference of the photon cylindrical inner structure. A wave property shown as the interference or the diffraction is in fact the statistical expression of assembly of monoenegetic self-entangled photons traveling along multiple individual trajectories and interacting aphysically with aphysical periodic structures such as multi-slit or crystal.

The position of the c-ring along the length of the aphysical cylinder is defined by the position parameter, PP, which is the quantum fundamental irreducible random "hidden" variable. For large assembly of monoenergetic photons the statistical density distribution of PP along the length of the aphysical energy cylinder is uniform. Such position parameter distribution in an assembly of photons explains the rhombus-like shape of the interference pattern in the Max-Zehnder interferometer. The interference pattern is the result of aphysical-aphysical interaction of the host and a-fraction. The c-ring physical energy plays no role in interference or diffraction.

Of all the elementary particles, the photon is one of the most versatile.

Mediating photons are real photons producing repelling (always repelling) action which explains "the charge screening" effect in case of electron-electron elastic interaction.

The photon spinning c-ring has enormous centrifugal force without centripetal force to balance.

The following is an explanation:

Traveling in free space, photon is precluded from expansion and release of its physical energy because there is no pathway available for such release.

As a general property of photon quantum ontology, a pair of photons (or several photons) can be entangled via a c-link connection. This pair remains entangled while traveling, with unlimited separating distance.

In the aphysical energy domain, photon acquires self-entanglement. The aphysical domain is the domain of aphysical-aphysical interaction only, such as no-loss optical material. In the aphysical domain, the photon aphysical energy cylinder splits on a number of a-fractions with only one, the host, carrying a c-ring with physical energy. Each a-fraction is connected to the host by its individual c-link. This quantum state is called self-entanglement, the massive phenomenon explaining photon interference and diffraction. In the aphysical domain, a-fractions and the host travel independently along their individual trajectories and participate independently in elastic aphysical–aphysical interactions with other quantum entities with no transfer or exchange of the aphysical energy between them.

In the optical domain, individual a-fractions can experience second order of self-entanglement, or even higher order of self-entanglement, via aphysical-aphysical interactions.

Under a certain relationship between periodicity of multi-slit structure and photon wavelength, interaction of the full-fledged photon with such multi-slit structure results in photon self-entanglement, where the host with intact physical energy c-ring travels through one slit and the other a-fractions travel through the other slits.

The interaction of optical photons with ideal optical material is not a physical process. Such process is governed exclusively by aphysical energy via aphysical-aphysical interaction. In a "no physical energy transfer environment", such as transparent optical materials in quantum optics, it is the aphysical substance of the photon, in the form of a set of a-fractions including the host, that plays the role in dynamics of quantum optics. The aphysical substance is the deterministic player in quantum optics with the exclusive role for interference and diffraction, there-

fore quantum optics should be called 'aphysical quantum optics.'

If an individual a-fraction of self-entangled photon attempts to engage in aphysical-physical interaction with another quantum entity, it is immediately recalled by the host. If the host meets such quantum entity, the host instantaneously recalls all a-fractions and transforms itself into the full-fledged photon.

When self-entangled photon encounters an absorbing physical entity such as a detector, it converts itself into the full-fledged photon and proceeds with complex physical-physical, physical-aphysical, and aphysical-aphysical interaction process such as

- *transforming its physical c-ring spin energy into physical kinetic energy and transferring it together with physical spin to the absorbing entity; and*
- *transforming its aphysical cylinder spin energy into aphysical kinetic energy column and transferring it together with aphysical spin to the absorbing entity.*

There is no resulted entanglement in photon-electron elastic and non-elastic interactions.

Another property of the photon known in quantum optics, is its ability, under certain physical-aphysical conditions, to split into secondary photons with all conservation laws applied. A known example is the down conversion, which produces pairs of correlated or entangled secondary photons. Nothing can prevent photon from further splitting.

♦ ♦ ♦

References

[1] P.A. Schilpp, ed., *Albert Einstein, Philosopher-Scientist,* MJF Books, New York, 1949, p. 672.

[2] Victor Vaguine, *Photon Non-physical Interactions and Quantum Enigmas,* International Journal of Modern Physics and Applications, vol. 1, No. 5, 2015, pp.219-223.

[3] Manjit Kumar, *Quantum: Einstein, Bohr and the Great Debate about the Nature of Reality,* W. W. Norton & Company, New York, London, 2010, p. 343.

[4] M. Jammer, *The Philosophy of Quantum Mechanics,* page 205, quoting W. Heisenberg, Physics and Philosophy, Allen and Unwin, London,1958, p. 160.

[5] Dipankar Home and Andrew Whitaker, *Einstein's Struggle with Quantum Theory; A Reappraisal,* Springer Science+Business Media, LLC, 2007, p. 62.

[6] *"...particles hover in quantum limbo, in a fuzzy, amorphous, probabilistic mixture of all possibilities; only when measured is one definite outcome selected from the many"* – Brian Greene, *The Fabric of the Cosmos: Space, Time, and the Texture of Reality,* Alfred A. Knopf, New York, 2004, p. 112.

[7] Albert Einstein in an interview with Alfred Stern, Contemporary Jewish Record 8, 245-9 (1945), also in A. Calaprice, ed. p. 253.

[8] Antoine Suarez, *Decision at the Beam-Splitter, or Decision at Detection, That is the Question,* arXiv:1204.5848v2 [quant-ph] 15 Apr 2013.

[9] Anton Zeilinger, *Dance of the Photons; From Einstein to Quantum Teleportation,* Farrar, Straus and Giroux, New York, 2010, pp. 80-81.

[10] Holger Gies, Felix Karbstein, Christian Kohlfurst, and Nico Seegert, *Photon-photon scattering at the high-intensity frontier,* arXiv: 1712.06450v1 [physics.optics], 15 Dec. 2017.

[11] C.K. Hong and L. Mandel, *Theory of parametric frequency down conversion of light,* Phys. Rev. A 31 (4), 2409-2418 (1985).

[12] P.G. Kwiat, K. Mattle, H. Weinfurter, and A. Zelinger, *New High-Intensity Source of Polarization-Entangled Photon-Pairs,* Phys. Ref. Lett. 75, 4337 (1995).

[13] Victor Vaguine, *Prologue to Super Quantum Mechanics*; ConsReality Press, 2012.

[14] E. J. Galvez and Mark Beck, *Quantum Optics Experiments with Single Photons for Undergraduate Laboratories,* Proc.SPIE 9665, Tenth International Topical Meeting on Education and Training in Optics and Photonics, 2007.

[15] Xian-Min Jin, Cheng-Zhi Peng, Tao Yang and Youjin Deng, *Interference of overlap-free entangled photons with a Mach-Zehnder-like interferometer,* arXiv: 1005.0802 v.1 [quant-ph] 5 May 2010.

[16] E.J. Galvez, C.H. Holbrow, M.J. Pysher, J.W. Martin, H. Courtmanche, L. Hellig, and J. Spenser, *Interference with correlated photons: Five quantum mechanics experiments for undergraduates,* Am. J. Physics 73 (2), 127, 2005.

Recommended Reading

1. P.C.W. Davies, J.R. Brown, (eds.), *The Ghost in the atom*, Cambridge University Press, 1993.

2. J.S. Bell, *Speakable and Unspeakable in Quantum Mechanics*, Cambridge University Press, 1991.

3. Dipankar Home and Andrew Whitaker, *Einstein's Struggles With Quantum Theory*, Springer, 2007.

4. James Evans and Alan S. Thorndike, *Quantum Mechanics at the Crossroads*, Springer, 2007.

5. A. Elitzur, S. Dolev, N. Kolenda, (ed. 2).

6. M. Schlosshauer, (ed.), *Elegance and Enigma; The Quantum Interviews*, Springer, 2011.

7. Bruce Rosenblum and Fred Kuttner, *Quantum Enigma; Physics Encounters Consciousness*, Oxford University Press, New York, 2006.

8. Anton Zeilinger, *Dance of the Photons. From Einstein to Quantum Teleportation*, Farrar, Straus and Giroux, New York, 2010.

9. Victor Vaguine, *Prologue to Super Quantum Mechanics*, ConsReality Press, 2012.

10. Mark Beck, *Quantum Mechanics; Theory and Experiment*, Oxford University Press, 2012.

About the author

Victor Vaguine, Ph. D., is the sole founder of Aphysical Quantum Mechanics, a deeper quantum theory and the origin of the Second Quantum Revolution.

Victor Vaguine is an independent scientist and philosopher.

He was born and educated in Russia (the former USSR). He earned his master's degree in Radiophysics (Electromagnetism) from the Gorki State University, Russia (1959).

From 1966 to 1971, he participated in the joint CERN-USSR program for development of equipment for high energy physics experiments at the European Center for Nuclear Research (CERN) in Geneva, Switzerland.

He received his Ph.D. in Physics from the University of Paris, Orsay, France (1970).

In 1971, after arriving to the United States, he accepted a scientist position at Varian Corporation, Palo Alto, California, an established high technology company specializing in radiation therapy equipment. A few years later, he was promoted to manager of a research & engineering division. He later left his secure position at Varian to take an active role in startup medical instrumentation companies.

As inventor and co-inventor, Vaguine holds 22 patents (United States and international). He has published numerous scientific papers and collaborated with several universities.

Since 2008 he has focused his scientific efforts on foundational issues of quantum mechanics and has developed Aphysical Quantum Mechanics, a deeper quantum theory, thus fulfilling Einstein's dream.

Aphysical Quantum Mechanics is the origin of the Second Quantum Revolution with immediate foundational transformation of quantum mechanics, quantum optics, and elementary particle physics. He has published his scientific work in three separate volumes under the title "*The Second Quantum Revolution*".

Victor Vaguine lives with his wife in Dallas, Texas USA.

His professional information can be found on *vaguine.com*

His email address is aphysical.qm@gmail.com.

Index

A

B

C

D

E

F

G

H

I

M

N

O

P

Q

R

S

T

U

V

Z

Made in the USA
Monee, IL
13 October 2020

44827882R00085